ARISTOTLE ON MATTER, FORM, AND MOVING CAUSES

This book examines an important area of Aristotle's philosophy: the generation of substances. While other changes presuppose the existence of a substance (Socrates grows taller), substantial generation results in something genuinely new that did not exist before (Socrates himself). The central argument of this book is that Aristotle defends a 'hylomorphic' model of substantial generation. In its most complete formulation, this model says that substantial generation involves three principles: (1) matter, which is the subject from which the change proceeds; (2) form, which is the end towards which the process advances; and (3) an efficient cause, which directs the process towards that form. By examining the development of this model across Aristotle's works, Devin Henry seeks to deepen our grasp of how the doctrine of hylomorphism – understood as a blueprint for thinking about the world – informs our understanding of the process by which new substances come into being.

DEVIN HENRY is an associate professor in the Department of Philosophy at the University of Western Ontario. He is co-editor of *Bridging the Gap between Aristotle's Science and Ethics* (with Karen M. Nielsen, Cambridge, 2015) and has published many articles on ancient Greek philosophy.

ARISTOTLE ON MATTER, FORM, AND MOVING CAUSES

The Hylomorphic Theory of Substantial Generation

DEVIN HENRY
University of Western Ontario

CAMBRIDGE
UNIVERSITY PRESS

CAMBRIDGE
UNIVERSITY PRESS

University Printing House, Cambridge CB2 8BS, United Kingdom

One Liberty Plaza, 20th Floor, New York, NY 10006, USA

477 Williamstown Road, Port Melbourne, VIC 3207, Australia

314–321, 3rd Floor, Plot 3, Splendor Forum, Jasola District Centre, New Delhi – 110025, India

79 Anson Road, #06–04/06, Singapore 079906

Cambridge University Press is part of the University of Cambridge.

It furthers the University's mission by disseminating knowledge in the pursuit of education, learning, and research at the highest international levels of excellence.

www.cambridge.org
Information on this title: www.cambridge.org/9781108475570
DOI: 10.1017/9781108646680

First published 2019

A catalogue record for this publication is available from the British Library.

Library of Congress Cataloging-in-Publication Data
NAMES: Henry, Devin, 1973– author.
TITLE: Aristotle on matter, form, and moving causes : the hylomorphic theory of substantial generation / Devin Henry.
DESCRIPTION: First. | New York : Cambridge University Press, 2019. | Includes bibliographical references and index.
IDENTIFIERS: LCCN 2019042624 (print) | LCCN 2019042625 (ebook) | ISBN 9781108475570 (hardback) | ISBN 9781108646680 (ebook)
SUBJECTS: LCSH: Aristotle. | Substance (Philosophy) | Matter. | Hylomorphism.
CLASSIFICATION: LCC B485 .H383 2019 (print) | LCC B485 (ebook) | DDC 110.92–dc23
LC record available at https://lccn.loc.gov/2019042624
LC ebook record available at https://lccn.loc.gov/2019042625

ISBN 978-1-108-47557-0 Hardback

For Christine, whose ousia *made this book possible.*
And for Evelyn and Jackson, whose genesis *almost made it impossible.*

Contents

Preface and Acknowledgements

There are a number of people that I need to thank for helping me bring this project to completion. First and foremost, I would like to thank my partner, Christine Sprengler, whose love, support, and ongoing inspiration made it possible for me to write this book. I am forever grateful for all she has done. On the academic side, I would like to express my gratitude to Mariska Leunissen, James Lennox, Michael Ferejohn, Mary Louise Gill, Jessica Gelber, Emily Kress, Adam Woodcox, and others who have commented on various chapters in one way or another. I am especially grateful to Sophia Connell, with whom I have had many correspondences. Her extensive comments on the manuscript ultimately helped me pull all of the material together into a more coherent argument. Her input had a profound effect on the final structure of the book. I am also indebted to Cecilia Li for her exceptional work preparing the subject index for the book. Finally, I would like to thank my two young children, Evelyn and Jackson, for choosing my cover image. I will always encourage your love of spiders.

Although I have discussed many of the themes in this book elsewhere, the present work is not just a collection of my earlier papers. Rather, it builds on my previous work and integrates it into a more comprehensive and unified picture of Aristotle's theory of substantial generation that not only goes beyond what I have said in the past but in some cases corrects it. The result is a more panoptic view of the terrain that one does not get from reading those individual papers. In what follows I attempt to walk the line between the Scylla of assuming too much familiarity with the secondary literature and the Charybdis of ignoring the work that has already been done. The former risks an account that is inaccessible and exclusionary, while the latter risks one that floats free of the relevant debates and controversies. Though I engage with the important literature throughout this book, I do so in a way that does not assume the reader is an expert in the various positions discussed therein. What follows thus aims to be as accessible as possible for a general audience with a basic understanding of the philosophy of Plato and Aristotle while at the same time acknowledging the exceptional work of my predecessors.

Abbreviations of Aristotle's Works

APo.	*Analytica posteriora (Posterior Analytics)*
DC	*De caelo (On the Heavens)*
DA	*De anima (On the Soul)*
De Juv.	*De juventute et senectute, De vita et morte (On Youth and Old Age, On Life and Death)*
NE	*Ethica Nicomachea (Nicomachean Ethics)*
GA	*De generatione animalium (Generation of Animals)*
GC	*De generatione et corruptione (Generation and Corruption)*
HA	*Historia animalium (History of Animals)*
IA	*De incessu animalium (Progression of Animals)*
MA	*De motu animalium (Movement of Animals)*
Metaph.	*Metaphysica (Metaphysics)*
Meteor.	*Meteorologica (Meteorology)*
PA	*Partibus animalium (Parts of Animals)*
Ph.	*Physica (Physics)*

All references to Aristotle's text follow Bekker pagination.
All references to Plato's text follow Stephanus pagination.
All translations are my own, unless otherwise specified.

Introduction

This book is an inquiry into an important yet neglected area of Aristotle's philosophy: the generation of substances. All change for Aristotle is a form of coming-to-be (or passing-away) in some sense. With some changes the substance already exists and simply becomes qualified in some way, as for example when Socrates (a substance) comes to be healthy from being unhealthy (a change in quality) or comes to be bigger from being smaller (a change in quantity). Aristotle calls this coming-to-be in the qualified sense (*gignesthai ti*). In cases of substantial generation (what Aristotle calls coming-to-be in the unqualified sense, *gignesthai haplôs*) the change results in a new substance coming into existence. At the end of development, we say that Socrates has come to be (full stop) rather than saying he has come to be *F*. This distinction was not universally accepted by Aristotle's predecessors. Some denied that generation involved anything genuinely new coming into existence that was not there before. Instead, they held that all putative cases of generation are simply the modification of some underlying element that remains numerically the same throughout the change, which is itself not subject to true generation or destruction. Part of Aristotle's challenge was thus to show that substantial generation is a distinct change in its own right that differs from changes in the non-substantial categories in some significant way.[1]

The central argument of this book is that Aristotle defended a "hylomorphic" model of substantial generation. In its most basic formulation this model says that new substances come into being from matter and form.[2] In *Generation of Animals* (*GA*) I 21 Aristotle says that in the case of animals the form comes from the male and the matter from the female

[1] See *GC* I 1 (discussed in Chapter 3).

[2] This book assumes a general familiarity with the basic concepts of Aristotle's philosophy (e.g. form and matter, substance, actuality and potentiality). An accessible introduction to the relevant parts of his system can be found in Lear (1988), Barnes (2000), Connell (2016, 9–12), and Shields (2016, xiv–xliv). For a good introduction to the relevant biological concepts see Peck (1942, xxxviii–lxx).

'so that one thing comes to be from them in the way that a bed is formed from the carpenter and the wood or the sphere from the shape and the wax' (729b15–18). Yet the claim that Aristotle defended a hylomorphic model of generation, familiar as it is to Aristotle scholars, leaves considerable room for interpretation. In this book I attempt to articulate, develop, and in some instances significantly revise the traditional interpretation of Aristotle's theory of generation with the aim of deepening our grasp on how the doctrine of hylomorphism – understood as a blueprint for thinking about the world – informs his understanding of the process by which new substances come into being.

As a way into the topic we can note that the very idea of a *genesis* of *ousia* combines two ontological categories that were typically opposed to one another in ancient Greek philosophy. The grand opposition between Being (*ousia*), on the one hand, and Coming-to-Be (*genesis*), on the other, has its roots in Presocratic philosophy. Parmenides saw being as the only metaphysical possibility, from which he deduced a number of shocking theorems including the claim that there is no change, no generation or destruction, and no differentiation among things; everything is everywhere full of being 'unmoved within the limits of great bonds' (Fr. 8). By contrast the famous Heraclitean dictum – 'Upon those that step into the same river different and different waters flow' (Fr. 12) – emphasised the idea of constant flux, that nothing ever *is* but everything is always *coming to be* (*Theaetetus* 152d2–e4). The opposition between being and becoming is at the core of Plato's metaphysics as well. The metaphysical distinction appears in the middle books of the *Republic*, where Plato presents his so-called Two-World view of reality:

> But our present discussion, on the other hand, shows that the power to learn is present in everyone's soul and that the instrument with which each learns is like an eye that cannot be turned around from darkness to light without turning the whole body. This instrument cannot be turned around from that which is coming to be without turning the whole soul until it is able to study that which is and the brightest thing that is, namely, the one we call the Good. (*Republic* VII, 518c3–d1, Reeve translation)

In Books VI and VII being is associated with the Forms, which are eternal and unchanging paradigms that serve as objects of understanding, while coming-to-be is associated with the changing world of sensible particulars that are at best objects of belief (*pistis*). This same Two-World view is articulated explicitly in the *Timaeus*:

> As I see it, then, we must begin by making the following distinction: What is that which always is and has no becoming, and what is that which is always coming to be and never is? The former is grasped by intellect together with an account and has everlasting being, while the latter is grasped by belief together with non-rational sensation, and it comes to be and passes away but never really is. Now everything that comes into being necessarily comes to be by the agency of some cause (for it is impossible for anything to come to be apart from a cause). So whenever the craftsman looks at what is always changeless and, using a thing of that kind as his model, reproduces its form and character, then, of necessity, all that he so completes is beautiful (*kalon*). But when he looks at a thing that has been begotten, his work will lack beauty. (27d5–28b4, translated after Zeyl)

According to Aristotle, Plato's whole metaphysical system was itself an attempt to reconcile these two aspects of reality in response to a problem about knowledge (*Metaph.* 987a29–b13). At *Theaetetus* 152c (Plato's) Socrates introduces two constraints on knowledge: (i) knowledge must be of what is; and (ii) it must be unerring. On Aristotle's interpretation Plato posited separately existing Forms to meet these conditions. He had accepted the Heraclitean view that the sensible world is in a constant state of becoming and so there cannot be any knowledge of it. And so, if knowledge is possible, it must concern an eternal, unchanging world of being (the Forms) that is separate from sensible particulars.

The distinction between being and coming-to-be was equally crucial for Aristotle's metaphysics. While he rejects the Two-World view of the *Republic* and *Timaeus* (the distinction does not correspond to a realm of separate Forms and a realm of sensible particulars), Aristotle does distinguish between a realm of ungenerated substances (God and the heavenly bodies) and a realm of sublunary substances that come to be and pass away (the elements, animals, and plants). What he denies is *the opposition* between 'that which always is and has no becoming' and 'what comes to be and passes away but never really is'. For on Aristotle's account, even though natural substances come to be and pass away, they are capable of partaking in everlasting being through perpetual generation (*GC* II 10, 336b33–5). On this view *the being* of natural substances is accomplished through their *coming-to-be*.[3]

[3] Quarantotto (2015, 185) calls this Aristotle's 'dynamic ontology', the idea that 'the world keeps being insofar as it keeps moving and changing'. As we shall see, Aristotle argues in *GC* II 10 that (pace Plato) generated substances are capable of partaking in everlasting being. In the case of living things, they do this by forming eternal lineages (*GA* II 1). I discuss this in detail in Chapter 9.

There is another aspect of the opposition between being and becoming that forms a major theme of this book (see especially Chapters 4 and 8). In the *Philebus*, Plato has Socrates argue for a certain teleological relation between them (though in this case without the same Two-World metaphysics in the background):

SOCRATES: There is, on the one hand, the *becoming* of all things, and on the other hand, their *being*... Now, which of the two do you think exists for the other's sake? Shall we say that coming-to-be occurs for the sake of being or does being exist for the sake of coming-to-be?

PROTARCHUS: Is what you want to know whether what is called being is what it is for the sake of coming-to-be?

S: Apparently.

P: By heavens, what a question to ask me! You might as well ask, 'Tell me, Protarchus, whether shipbuilding occurs for the sake of ships or whether ships are for the sake of shipbuilding', or some such thing.

S: That is precisely what I am talking about, Protarchus ... I hold that all the means of production, as well as all tools, and quite generally the materials are provided for the sake of some process of generation. I further hold that every process of generation in turn always takes place for the sake of some particular being and that all generation taken together occurs for the sake of being as a whole.

(*Philebus* 54a5–c4)

On this account being is the end for the sake of which coming-to-be occurs rather than vice versa. For example, everything that has to do with the process of shipbuilding exists for the sake of the ship that is built and not the ship for the sake of these.[4]

Aristotle follows the *Philebus* in holding that coming-to-be occurs for the sake of being rather than vice versa. This commitment shaped his account of substantial generation and his views on how we should go about investigating it. For example, in the first book of the *Parts of Animals* (*PA*) he writes:

> We should also not forget to ask whether it is appropriate to state, as those who studied nature before us did, how each thing has naturally come to be, rather than how it is. For the one differs not a little from the other ... For even with house-building, it is more the case that these things happen

[4] Socrates goes on to associate the good of each thing with the end for the sake of which it comes to be, while generation and everything that has to do with it 'belongs in another class' (54c10–d3). The main function of the argument is to show that pleasure has no intrinsic value: since pleasure 'is always a process of becoming and that there is no being at all of pleasure' (53c5–8; 54c7–8), it follows that pleasure is not among the things that are good in themselves. Compare the argument against hedonism at *Gorgias* 499b–500b.

> because the form of the house is such-and-such than that the house is such-and-such because it comes to be in this way. For coming-to-be is for the sake of being. (*PA* I 1, 640a10–18, Lennox translation with modifications)

He echoes this point at the beginning of *GA* V:

> In the case of all those ordered and determinate products of nature, it is not because each thing *comes to be* of a certain character that the thing *is* of a certain character but rather it is because they *are* such-and-such that they *come to be* such-and-such. For generation (*genesis*) follows on substantial being (*ousia*) and occurs for the sake of it, but not this on generation. (*GA* V 1, 778b3–6)

In this context *ousia* (being) refers to the way a thing exists (640a12: *pôs estin*) when it has finished coming into being and has reached a state of completion (*to entelecheia*). In Aristotle's metaphysics the form of a thing constitutes its *ousia* insofar as its presence in the matter is what causes it *to be* a substance of such-and-such a kind (*Metaph.* 1041b5–9). To say that *genesis* occurs for the sake of *ousia* is thus just to say that the process of generation aims at the form that constitutes the substantial being of the thing that is coming to be and thus determines that process's structure.[5] For example, the being of a house is given by its function (e.g. providing shelter), which is the form specified in the definition stating what it is to be a house. This form structures the process of house-building (as a goal) insofar as it determines what sorts of materials must be used and what sorts of changes they must undergo in order to produce that form in those materials (*PA* 639b21–30).[6] Likewise in the case of naturally generated substances. While it is true that wings and feathers are the result of certain changes that occur during the process of a bird's development, Aristotle holds that those changes themselves ultimately occur in the way that they do *because* what is coming into being is a bird whose actualised form includes the capacity for flight which requires ('conditionally necessitates') those kind of structures. It is in this sense that coming-to-be 'follows on and occurs for the sake of' being rather than vice versa.[7]

The priority of being to coming-to-be also has methodological implications. Since generation occurs for the sake of the substantial being it

[5] Lennox (2001c, 181).

[6] See also *PA* 646a24–b9, where Aristotle puts the point in terms of being "prior in account". Compare *GC* II 9, 335b33–5 (discussed in Chapter 4).

[7] At *PA* 641b30–1 Aristotle extends this to the relation between an embryo and the substance that comes to be from it: 'But prior even to the embryo is that of which it is an embryo; for while the embryo is a *genesis* the end is an *ousia*.'

produces, our understanding of that process is dependent on our grasp of the form of the thing that is coming to be. As Aristotle says at *PA* 640b1–3, 'since it *is* such, its generation necessarily happens in this way and is such as it is; this is why one part comes to be first, then another'. For example, we first need to determine what a fish is, what its parts are for, and what sort of structure they must have in order to execute those functions before going on to investigate the process of its development. This helps to explain the ordering of the *Parts of Animals* relative to the *Generation of Animals*. Since coming-to-be is for the sake of being, it follows that explanations of the sorts we find in the *PA* (which ask why animals have the parts they do) are explanatorily prior to the account of development we find in the *GA*.[8]

Finally, I shall argue that matter and form themselves play different causal roles depending on whether we are talking about the *being* or *coming-to-be* of natural substances. For example, the form of an *F* serves both as a formal cause of *being* an *F* – it is that feature of a thing whose presence in the matter causes it *to be* an *F* – and as a final cause of *becoming* an *F* – not by being present in the matter (the form is not present until the end of the change) but by being the goal *towards which* that process is directed. In this way the form tells us both what it is to be an *F* and why the generation of an *F* occurs in the way that it does (why the change follows the pattern it does). This difference in context will be important for thinking about Aristotle's hylomorphic account of substantial generation.

While the scholarship on Aristotle offers countless studies of his theory of substantial being, especially as it is developed in the central books of the *Metaphysics*, his account of substantial generation remains relatively neglected. The current study aims to address that imparity.[9] However,

[8] Lennox (2001c, ix). Cf. *PA* I 1, 640b1–3.

[9] Gill (1989) is a notable exception to this, and I have learned much from her exceptional study (as will become clear in what follows). Like my own project, she attempts to provide a systematic account of Aristotle's views on substantial generation that draws on his biological works, specifically the *Generation of Animals*. However, her book is still aimed at providing an interpretation of the doctrine of substantial being in the central books of the *Metaphysics*. Gill's study also reflects a tendency among scholars to mine the biological works for insights into Aristotle's other philosophical doctrines rather than thinking about them as containing philosophical insights of their own that 'might be capable of changing our frame of reference' (Connell 2001, 299). The other notable exception is Connell (2016), whose masterful study of the female in Aristotle's *Generation of Animals* provides a thorough account of Aristotle's biological views on generation. My own book falls somewhere between these two. Unlike Connell's book, my focus is on understanding Aristotle's philosophical theory of substantial generation more broadly and not simply on providing an interpretation of his biological theory of animal generation and the role of the female in that account (e.g. Chapters 1 to 4 deal with *Physics* and *Generation and Corruption*). Unlike Gill's book, the goal of my inquiry is to understand Aristotle's views on substantial generation independently of what it can tell us about his doctrine of substance. And so my project does not approach the biological works

before introducing the main argument of the book let me first say more about what counts as "substantial generation". This will help to better circumscribe the scope of my project. For this book is not a study of Aristotle's theory of change more broadly but only that kind of change that results in a substance coming into being.

One of the most fundamental distinctions in Aristotle's natural science is the distinction between coming-to-be in the qualified sense (*gignesthai ti*) and coming-to-be in the strict or unqualified sense (*gignesthai haplôs*).[10] All change for Aristotle is a form of coming-to-be (or passing-away) in one of these two senses. We can start by drawing the distinction linguistically by noting that the phrase "*S* comes to be" has a complete and incomplete form. When used in the qualified sense "*S* comes to be" remains incomplete without specifying what the subject comes to be ("*S* comes to be *F*"), as when we say that a man comes to be *healthy* or comes to be *bigger*. When used in the unqualified sense "*S* comes to be" is complete all on its own; we say that *S* comes into being, full stop. For example at the end of development we say that a human has come into being that did not exist previously.[11] Aristotle thinks the distinction also has an ontological dimension that turns on which category the change concerns.[12] When a thing comes to be in the qualified sense the change always occurs with respect to one of the non-substantial categories: alteration is coming-to-be in the

simply as a resource for shedding light on some other set of core philosophical doctrines. Instead, Chapters 5 to 8 examine the *Generation of Animals* in its own right, as a central part of the story I wish to tell about the development of Aristotle's hylomorphic model of substantial generation. At the same time, it will also be important to grasp the *GA* account within the broader context of Aristotle's system as a whole as well as in its historical context (e.g. by examining its roots in Plato). Only then will we be able to appreciate the complete picture of Aristotle's hylomorphic model of generation and its application to living things.

10 See, for example, *Ph.* 186a14–15, 190a31–3, 225a12–17, *GC* 318a31–5, 319a3–4, *Metaph.* 1042b7–8.

11 Cf. *APo.* 89b32–90a14. There is an ongoing scholarly debate surrounding the use of the Greek verb "to be" (*einai*) and, by extension, "comes to be" (*gignesthai*). For a discussion of the issues at stake see Kahn (2009) and Brown (1994, 1986). Simplifying as much as possible, the debate concerns whether or not the Greeks recognised a complete use of "is" (the syntactic issue) and, if so, what the precise meaning of this "is" is (the semantic issue). In the latter case, scholars have traditionally assumed that the complete use of the verb, "X is", has an existential sense ("X exists"), while others have insisted that it also has a veridical sense ("X is true" or "X is the case") (Kahn 1981, 1986). In this book I shall assume that Aristotle recognised a complete use of the verb "comes to be" (which corresponds to his unqualified use of the term) and that in this context it means "comes into existence". By contrast, the so-called predicative (or qualified) use of the verb ("X come to be F") describes a situation where X already exists and comes to acquire an attributed of some kind (e.g. X comes to be healthy or red).

12 Aristotle draws the distinction along categorial lines at *GC* I 3, 319a11–14, where he says that qualified and unqualified becoming can be distinguished 'in accordance with the categories' (*tais katêgoriais*); 'for all those things that do not signify substance are not said to come into being without qualification but only in the qualified sense'. And at *Physics* I 7, 190a32–3 he says that *only* substances come to be without qualification. See Philoponus, *in GC* 8,29 and Joachim (1926, 88).

category of quality (e.g. Socrates comes to be healthy); growth is coming-to-be in the category of quantity (e.g. Socrates comes to be bigger); locomotion is coming-to-be in the category of place (e.g. Socrates comes to be in the market).[13] All of these count as instances of coming-to-be in the qualified sense because the substance (a man) already exists and simply becomes qualified in some particular way. Coming-to-be in the unqualified sense, by contrast, is characterised as a change in the category of substance; it is *substantial* generation. Understood in this way unqualified becoming differs from qualified becoming in that the former, but not the latter, results in a new substance coming to be that was not there before. In this book I shall reserve the term "generation" for coming-to-be in the unqualified (*haplôs*) and strict (*kuriôs*) sense, while "coming-to-be" without any further specification will be used more broadly to include all forms of change (including alteration, growth, and locomotion).

While Aristotle maintains a sharp distinction between qualified and unqualified becoming throughout the corpus, the way he distinguishes them (along categorial lines) immediately raises a problem. What is at stake here is precisely what counts as *substantial* generation. The problem I have in mind is whether the coming-to-be of the simple bodies (elemental transformation) counts as substantial generation or whether Aristotle means to restrict the latter change to the generation of living things. The problem comes in the form of a trilemma. Aristotle insists that only substances come to be without qualification (e.g. *Ph.* I 7, 190a32–3; *GC* I 3, 319a11–14). And in *GC* he uses elemental transformation as a paradigm for that sort of change (e.g. *GC* I 4), which entails that the elements are substances. And yet in *Metaphysics* VII 16 he denies that the elements are substances on the grounds that they lack the requisite unity (1040b8–10). By the end of *Metaphysics* VII living things alone emerge as genuine (composite) substances. But if only living things count as substances, then only living things can be said to come into being without qualification. These three commitments form an inconsistent triad:

(1) Only substances come to be in the unqualified sense.
(2) The elements come to be in the unqualified sense.
(3) The elements are not substances.

The problem is obvious. (1) and (2) imply the denial of (3), (2) and (3) imply the denial of (1), while (1) and (3) imply the denial of (2). In order to resolve the tension Aristotle needs to deny at least one of these three

[13] *GC* 319b31–3.

propositions. This is no easy task since he seems to be committed to each of them.

One way to resolve the tension is to qualify (3). For example, in *Metaphysics* VII 16 Aristotle treats the elements as *potential* substances insofar as some particular substance might come to be from them. But this does not help. As we shall see in Chapter 1, Aristotle's settled view is that coming-to-be without qualification proceeds from a *potential* substance to an *actual* substance. If no element counts as an actual substance, then as long as (1) is true we cannot say that (e.g.) air comes to be from fire without qualification. For at best that change results in something that is a substance only potentially.

Another option is to accept (1) and (3) but reject (2). For example, on one reading of Aristotle, prime matter serves as a persisting subject that underlies every elemental transformation.[14] This makes elemental transformation a form of *qualified* becoming, since (at least according to the account in *GC* I 1–4) Aristotle calls it 'being altered' whenever the underlying subject remains through the change (e.g. 314b1–6).[15] This strikes me as the least attractive option. For Aristotle goes to great lengths in *GC* I 4 to defend the view that elemental transformation is a form of substantial generation and *not* an alteration of some persisting subject. More importantly, the very model of generation developed in *GC* is built up almost entirely from considerations of the elements and their mutual transformations.[16] So denying (2) would wreak major havoc on the *GC* account of coming-to-be.

A third solution would be to accept (2) and (3) but deny (1). This is an interesting alternative. On this reading unqualified becoming and substantial generation come apart, which leaves room for saying that other things besides substances come into being without qualification. This would allow Aristotle to treat elemental transformation as a paradigm case of coming-to-be in the unqualified sense (in accordance with *GC*) while maintaining that living things alone count as genuine substances (in accordance with the *Metaphysics*). While the elements may come to be

[14] Scholars have traditionally assumed that Aristotle thinks all change requires some persisting item that remains as a constituent of the finished product. Many take prime matter to play this role in substantial generation (e.g. Robinson 1974, 168). On the traditional view, prime matter is 'a bare "stuff", lacking all positive determinations, which is the matter of the elements and which makes elemental change possible'. It is normally understood as 'pure potentiality' that can only exist in the form of some determinate (or proximate) matter, e.g. as fire or earth (Robinson 1974). This book denies that Aristotle was committed to any of these claims.

[15] See Chapter 3.

[16] See Joachim (1926, pp. xxxvi–xxxvii), Burnyeat (2004, 7), Brunschwig (2004), Sedley (2004).

without qualification, only biological generation counts as substantial generation because it is only in the case of living things that unqualified becoming results in the emergence of a new *substance*. But this is not a live option either. While Aristotle does make a distinction in *GC* I 3 between elemental transformation and other forms of unqualified becoming (319a3–11), that distinction does not turn on whether or not what comes to be is a substance. By contrast, as we have seen, Aristotle routinely aligns coming-to-be *haplôs* with the category of substance (see note 12 above). There is simply no good reason to think that he means to treat substantial generation only as one species of unqualified becoming.

A more drastic approach to the trilemma is to accept all three propositions and then explain away the tension by claiming that Aristotle held propositions (2) and (3) at different points in his career. This way of resolving the tension is not completely unmotivated since it is possible to argue that Aristotle changed his mind about the ontological status of the elements. For example, at the beginning of *De caelo* III (a work with close ties to *GC*) Aristotle includes the elements on a list of natural substances:

> Since of the things that exist by nature we call some substances (*ousiai*), others functions, and others attributes of substances – by "substances" I mean the simple bodies (fire, earth, and the other terms of the series), and all things composed of them, including the heaven as a whole and each of its parts, as well as animals and plants and their parts; by "attributes" and "functions" I mean the movements of these and of all other things in which they have a power in themselves to cause movement, and also their alterations and reciprocal transformations – it is clear that the greater part of the inquiry into nature concerns bodies. For all natural substances (*hai phusikai ousiai*) can be classified either as a body or a thing that comes to be together with body and magnitude. (*DC* III 1, 298a27–b2)

Aristotle is explicit in this passage that the elements count as natural substances (cf. *Metaph.* V 8). Interestingly, in *Metaphysics* VII 2 Aristotle presents this exact list as a preliminary account of things that *are believed* to be substances (1028b8–13). When he returns to the list in *Metaphysics* VII 16 the elements are disqualified on the grounds that none of them is a unity but each one is 'like a heap' (1040b8–10). In light of this it would not be unreasonable to suggest that Aristotle simply changed his mind on the ontological status of the elements. This would resolve the tension because (2) and (3) would represent different phases in Aristotle's thinking about substance. In *GC* he treats the elements as natural substances (as he does in *DC*), and so it would have been perfectly reasonable for him to use elemental transformation as his model for coming-to-be without qualification. When he later came to write the central books of the *Metaphysics* he reconsidered the *DC* list and decided that the elements do not meet the strict requirements for being a substance, and so (by implication) their transformations no longer count as cases of coming-to-be without qualification since only substances come into being in that sense. By the end of *Metaphysics* VII living things alone are left standing as candidates for (composite) substances, and so only living things can be said to come into being without qualification.[17] Elemental transformation would then be downgraded to an alteration of some underlying subject (whether prime matter or something else).

The downside to this approach is that it is difficult to prove that Aristotle changed his mind on the ontological status of the elements in the absence of any confession to such a change. Fortunately, there is another way to resolve the tension that is consistent with the evidence but does not require such a radical solution.[18] What I have in mind is a variation on the fourth approach. Rather than saying that Aristotle held (2) and (3) at different stages in his career, a more promising solution is to take him to be working with a different concept of substance in (1) and (3).

One of the ways that Aristotle uses the concept of substance, especially outside the central books of the *Metaphysics*, is for what he calls a *tode ti*: a separately existing individual "this" that serves as an ultimate subject of predication.[19] This is clearly how Aristotle intends to use the concept in *GC*. For he speaks of being a "this", being "separate", and being a "substance" interchangeably throughout the treatise (see, e.g., *GC* 317b8–11, b31–2). It is likely that this is also the sense of substance that

[17] Of course, not all living things for Aristotle are subject to generation and destruction. God and the heavenly bodies are alive but neither come to be and pass away (e.g. *Metaph.* XII 2, 1069b25–6). My discussion of living things is restricted to ordinary plants and animals.

[18] Cf. Ferejohn (2013, 18–20). Various themes discussed in this book raise questions about the relation of Aristotle's different writings on substantial generation. While I take Aristotle to be a 'systems builder' (Barnes 1982, 36), I reject any kind of strong unitarian interpretation according to which all of his various discussions of substantial generation across the different works aim at presenting a unified and unchanging doctrine. Instead, I am sympathetic to Ferejohn's views on the exploratory nature of Aristotle's investigations (2013, 20). Thus, in this book I allow for a weak form of developmentalism where Aristotle's thinking about substantial generation develops from one treatise to another but not in a way that implies a significant change in philosophical commitments. Instead, the developments reflect a move from an *unreflective* position where Aristotle is speaking loosely or in an unreflective way about *X* to a more *considered* position where he is carefully reflecting on the nature of *X*. What I mean by this will become clear as the argument of this book proceeds.

[19] See, for example, *APo.* I 4, 73b7–8 and *Metaph.* V 8, 1017b13–14. In both places the concepts of being a substance, being a subject of predication, and existing in its own right (i.e. separately) are all interconnected. Aristotle explicitly invokes this concept of substance at *GC* 317b18–22 ('whether substance, in the sense of a "this something", comes to be at all').

Aristotle has in mind in the *DC* passage when he includes the elements among the list of natural substances (cf. *Metaph.* 1017b10–14), which he uses interchangeably with "body" (*soma*). By contrast, while *Metaphysics* VII also sees a substance as a *tode ti*, by the time Aristotle gets to *Metaphysics* VII 16 this is no longer treated as sufficient for substancehood. By that point in the inquiry, substancehood also requires the kind of unity that Aristotle thinks is missing in the case of the elements, which are mere heaps (1040b8–10).

There are good reasons for thinking that *Physics* I is using substance in this weaker sense, too. In *Physics* I Aristotle is largely engaged with the views of his predecessors, who do not share his metaphysically loaded sense of *ousia* from the central books of the *Metaphysics*. In that case, when Aristotle says in *Physics* I 7 that only substances come to be without qualification, we can take him to be talking about substances in the non-technical sense of separately existing "thises". And that is all he needs to say. For all he means to rule out is the possibility that accidental unities (so-called kooky objects[20]) come to be without qualification. For example, we do not say that the pale man comes to be without qualification but only that a man comes to be *pale*. Or to use Aristotle's example, we say the student comes to be *learned* not that it comes to be full stop (*GC* 318a34–5, 319a3–16).

We can now return to the three propositions that generated the inconsistent triad:

(1) Only substances come to be in the unqualified sense.
(2) The elements come to be in the unqualified sense.
(3) The elements are not substances.

My way of resolving the tension between these three propositions is to argue that the concept of "substance" is not being used in the same way throughout. When Aristotle says that only substances come to be in the unqualified sense, he means substance in the sense of a separately existing individual (or *tode ti*) that serves as an ultimate subject of predication. The elements are substances in that sense, and so they should count among those things that come to be without qualification.[21] When

[20] The term "kooky objects" was coined by Matthews (1982).

[21] Aristotle clearly thinks the elements are separately existing "thises" (*Metaph.* V 8). In *Metaph.* VII 16, he disqualifies the parts of organisms as candidates for substance on the grounds that they are not separable, that is, they are not capable of existing in their own right but only exist (as the functionally defined entities they are) when attached to a living body (1040b5–8). The elements are then ruled out, not because they are incapable of separate existence, but because they are not unified in the right way (b8–10). This suggests that Aristotle still thinks the elements, unlike the parts of animals, enjoy separability.

Aristotle denies that the elements are substances in *Metaphysics* VII 16, he means substance in the more robust sense that also requires being a unity. The elements are not substances in that sense since they lack the requisite unity (1040b8–10). But Aristotle nowhere commits himself to the claim that only substances *in that more robust sense* come into being without qualification.

Although I shall have things to say about the elements in what follows, in this book I focus mainly on the generation of biological substances for two reasons. First, by focusing on organisms I am hedging my bets. As we have just seen, it isn't obvious that elemental transformation counts as a form of substantial generation. Aristotle insists that only substances come to be without qualification. And it remains controversial whether he thinks of the elements as substances in the relevant sense. While I have tried to make the case that he does, if it turns out that the elements are not substances, then it will be false to say that when one element comes to be from another the change results in a new substance that was not there before.[22] Moreover, even if the elements do turn out to be substances, scholars do not agree on whether the process of elemental transformation itself conforms to Aristotle's model of substantial generation. *Physics* I 7 holds that all naturally generated substances come to be from matter, form, and privation (Chapter 2). Most scholars assume that within this model matter serves as the persistent subject that survives to become a constituent of the thing at the end of the change in the way that bronze survives in the finished statue. And yet it is not clear that anything plays this role in cases of elemental transformation.[23] While I shall try to show that these cases do conform to Aristotle's basic model (albeit in a limited way), it is universally agreed that his analysis does apply to the generation of those 'metaphysically uncontroversial substantial beings like plants and animals' (Burnyeat 2004, 12).

Second, and more importantly, even if one agrees that the elements do count as genuine substances and that their transformations do count as

[22] Gill (1989, e.g. 65) also sees elemental transformation as a controversial case that differs strikingly from ordinary cases of substantial generation. She attributes to Aristotle what she calls the 'Construction Model' of substantial generation, which crucially involves an increase in complexity (e.g. the frog is more complex than the egg from which it develops). In elemental transformation, however, a simple body changes into another simple body. But I don't see this emphasis on increased complexity (which Gill finds in *Metaphysics* VIII 5) as an essential part of Aristotle's account of substantial generation.

[23] See Krizan (2013) and references therein. I return to this puzzle at the end of Chapter 3. I argue that elemental transformation conforms to the basic hylomorphic model from *Physics* I 7 if we give up the idea that matter persists in cases of substantial generation.

substantial generation, there are economical reasons for focusing on the generation of living things. A detailed study of every form of substantial generation in the sublunary world would be an enormous project that would require a much longer book. In addition to a more thorough examination of elemental transformation than I offer, one would also have to sort through the details of Aristotle's account of the generation of uniform substances in *Meteorologica* Book IV. For he is explicit there that this change, too, counts as a form of coming-to-be in the unqualified sense (Chapter 4). Since Aristotle treats living things as paradigm natural substances (*Metaph.* 1034a3–4), biological generation turns out to be substantial generation par excellence. And so, we should expect all of the principles of substantial generation to apply to that phenomenon. By contrast, those other cases lack some important features that are exemplified by the generation of living things. For example, they do not exhibit goal-causation: neither the mutual transformation of the elements nor the generation of uniform substances can be said to occur *for the sake of* promoting the good condition of the thing that comes to be from them.[24] So focusing on biological generation allows us to capture certain key aspects of the change that would otherwise be missed if we were to focus exclusively on those other phenomena.[25]

The basic argument of this book is that Aristotle employs a single model of generation throughout the corpus – the hylomorphic model – which specifies the 'principles and causes' from which naturally generated substances originally come into being (e.g. *Ph.* 190b17–20, *Metaph.* 1032a12–14, 1033a24–8, 1069b35–1070a3).[26] According to the basic model introduced in *Physics* I 7, matter plays the role of the underlying subject (*to hupokeimenon*) from which a thing comes to be non-accidentally, while form is one of a pair of opposites, namely, the positive state acquired as a result of the change. I examine this model in Chapters 2 and 3. On the standard reading, *Physics* I 7 also holds that something must persist through the change and that in cases of substantial generation this role is played by

[24] For the most part the *Meteorologica* IV account is devoid of any reference to goal-causation. While the last chapter suggests that teleology is present even in the case of the uniform substances, I agree with Gill (2014) that what Aristotle has in mind is their formation within the context of living things (e.g. flesh and bone).

[25] Compare Lennox (2004b, 63).

[26] Aristotle's natural science distinguishes natural substances that come to be and pass away (so-called sublunary substances: the elements, plants and animals) from those that are not subject to generation or destruction (the heavenly bodies). In this book "substantial generation" and "natural generation" will be used interchangeably to refer to the generation of the former (natural *sublunary* substances).

matter. On this reading matter is made to do 'double duty' (Bostock 2006, 9) as both the starting point of the change and as what persists through it. This reading has been tremendously influential in the secondary literature, and I return to the theme of the persistence of matter throughout the book. In contrast to the orthodox view, I shall argue that matter's role in substantial generation is restricted to that of a pre-existing subject from which the change proceeds (the *terminus a quo*). Aristotle does not mean to commit himself in *Physics* I 7 (or anywhere else) to the further claim that matter also survives intact to become a constituent of the substance that comes to be. Indeed, as I go on to show in Chapter 3, Aristotle positively rejects that claim in *Generation and Corruption* I. On the reading I defend, this is what ultimately separates generation from changes in the non-substantial categories.[27] Because the idea of persistent matter is so entrenched in the literature, I will need to spend considerable time trying to loosen its grip.

When Aristotle returns to the principles of natural generation in *GC* II 9, he expands this basic model in three important ways (see Chapter 4). First, matter is conceptualised there as the seat of a thing's 'capacity to be and not be' (335a32–3: *to dunaton einai kai mê einai*). I argue that this is the essence of Aristotelian matter in the sense of its causally basic feature. For the fact that matter is a capacity of this sort explains why it is suitable to play the role of subject of generation (and destruction) in the first place.[28] The role of form in generation is also recast in *GC* II 9 as *that for the sake of which* the change takes place and not simply as the positive state *into which* a thing develops (the *terminus ad quem*). This makes form a limit (*peras*) in the normative sense of the end or goal towards which the process is directed (*Metaph.* V 17, 1022a4–13).[29] Finally, Aristotle goes to great lengths in *GC* II 9 to show that these two principles alone are not sufficient to account for generation; instead, we must also have recourse to a third principle over and above these. By drawing on a parallel passage in *Metaphysics* I 3, I argue that what Aristotle is after here is a goal-directed efficient cause that can explain why

27 This interpretation removes the standard motivation for introducing a concept of prime matter (see note 13). On my reading, there is no special need for something like prime matter, because Aristotle does not think the underlying subject persists even in elemental transformation. This differs from the strategy Robinson (1974) attacks, which accepts the need for a persisting subject and then looks around for something to play that role.

28 Contrast Ebrey (2018), who argues that matter's most fundamental feature is being a subject of generation. On my reading this feature follows from matter's essence as a capacity to be and not be.

29 *GC* 335b6–7. *Physics* I 5–7 characterises form merely as a limit of change but not in the sense of a goal.

things come to be in a 'good and beautiful condition' (*eu kai kalôs*) (984b8–22).

This completes the first major aim of the book, which is to develop an account of Aristotle's hylomorphic theory of substantial generation in its abstract form. According to this account the principles from which naturally generated substance originally come to be are threefold: (1) matter, which is the subject from which the change proceeds; (2) form, which is the end towards which the process advances; and (3) a goal-directed efficient cause, which is the agent that directs the process towards that form. The other major aim of the book is to build towards an account of substantial generation in the biological realm, which is substantial generation par excellence. Examining Aristotle's account of that change is the project of Chapters 5 to 8, which focus mainly on the theory articulated in the *GA*.

In Chapters 5 and 6 I consider how the abstract hylomorphic model examined in the first half of the book applies to cases of animal generation. At the core of what I call Aristotle's "reproductive hylomorphism" is the claim that male and female are principles (*archai*) of generation: the male is the efficient cause that supplies the form, while the female supplies the matter that receives that form. This provides the general model, which Aristotle applies to two different aspects of animal generation. I argue that in *GA* Book I Aristotle uses this model as a way to analyse the causal contributions that male and female seed (*sperma*) make to embryogenesis. In this context menstrual blood serves as the matter for the change, while semen gives form to this matter. The product of their interaction is an embryo, which then develops into a new animal after being catalysed by the heat in the male semen.[30] In Book II Aristotle uses the hylomorphic model as a way to analyse the contributions that male and female animals make to the completed substance itself. In this context Aristotle says that the form supplied by the father is the offspring's soul, while the mother provides the body that serves as the underlying subject of that form.

[30] In this book I draw on a distinction between embryogenesis and morphogenesis. I use "embryogenesis" to refer to the initial formation of the embryo, which includes what we call fertilisation (what Aristotle calls "setting" the embryo) as well as the formation of its primitive heart. Cf. *HA* VI 3, 561a6–18. "Morphogenesis" refers to the subsequent stage of development in which the new animal comes to acquire its characteristic shape and form as the rest of its parts are differentiated out of the matter. See *GA* 776a33–4 where Aristotle talks about the stage where the matter is 'being expended on moulding the fetus' (*plasin tou embruou*). *GA* II 4, 739b33–7 marks an important transition in the *GA* theory between the account of embryogenesis and the account of morphogenesis (cf. 741b25–6). Aristotle does not explicitly invoke the hylomorphic model in the latter case, though I shall argue that here too we can identify a pre-existing matter for the change (the embryo) as well as an efficient cause that imposes form on that matter (the offspring's own nature or soul). See Chapter 7.

Aristotle goes on in *GA* II 5 to raise a puzzle about the role of males in generation: Since the female provides both the matter and place for generation, why is there a need for males in addition? Why can't the female generate an animal on her own? His answer is that, while the female may provide nutritive capacities of soul, the male alone provides the capacity for sensation. And since animals are defined by sensation, even if the female were able to generate a living thing by herself, what she would generate would not be *an animal* by definition. For it would lack the defining feature in virtue of which something is an animal. I argue that the fact Aristotle even raises this puzzle shows that the hylomorphic model presented *GA* II 4, which initially divides parental contributions along body–soul lines, cannot be the whole story. On the interpretation I defend, the argument in *GA* II 5 is intended as a further development of this more general model: *GA* II 4 first tells us that the male provides soul, which is 'the substantial being of a certain kind of body' (738b18–27); *GA* II 5 then goes on to circumscribe this more precisely by saying the male provides *the sensory part* of the soul, which is the substantial being of the *animal* body. This is the substantive conclusion of Chapter 6. If form (*eidos*) is taken in the narrow sense to refer to essence, then there is a sense in which the father alone provides the animal's form insofar as he provides the sensory part of its soul (which is that feature that makes a living thing an animal). But if form is taken in the broader sense to pick out the soul in general (*DA* II 1), then both males and females can equally be said to make a formal contribution to the substance that comes to be from them.

Chapter 7 explores a different puzzle about the efficient cause of animal generation. On the one hand, Aristotle defines naturally generated substances as those whose generation is derived from an *inner* principle of change, i.e. its nature (*Ph.* II 1). Most scholars take nature in this sense to be operating as an internal efficient cause, and so we should expect animal generation to be the type of change that is efficiently caused by the nature in the embryo itself. On the other hand, Aristotle's reproductive hylomorphism identifies the father as an external efficient cause, who forms the offspring using his semen as a tool (e.g. *GA* 730b21–4, 734b36–735a4). But in that case, it is hard to see how animal generation qualifies as a natural change at all. The central aim of Chapter 7 is to reconcile these two alternative perspectives on the primary efficient cause of animal generation.

In Chapter 8 I turn my attention to Aristotle's account of morphogenesis and ask how the three principles of the hylomorphic model – matter, form, and moving cause – interact in the dynamic teleological context of organic development (the process where the matter gradually comes to acquire a new substantial form). In *GC* II 9 Aristotle considers a version of materialism according to which natural generation is the effect of certain physical forces rooted in the elements out of which everything comes to be: 'For since, as they say, heat naturally divides things and cold naturally amalgamates them, and that of the rest of the contraries some act while others suffer action, it is out of such things [as matter] and by their agency [as efficient cause] that everything else comes to be and passes away' (*GC* 336a3–6). Aristotle objects to this theory on the grounds that the efficient causes invoked by the materialists (e.g. fire, heat and cold) are at best proximate causes of generation that are used by the primary efficient cause as "tools" for producing more complex ends (336a1–12). Aristotle's objection in *GC* II 9 is brief and not clearly expressed (see Chapter 4). In particular, we are left wondering how the actions of the primary efficient cause are related to those elemental forces that he agrees are at work in natural generation. Chapter 8 attempts to flesh out this account by drawing on various things Aristotle says in his biological works. According to the picture that emerges from the biology, all the "hands-on" labour of development is assigned to the activity of certain lower-level elemental forces (such as heating and cooling) whose necessary causal interactions physically transform the embryo into a new substantial individual. The "choreography" of those interactions (their timing, location, order, etc.) is due to some other goal-directed efficient cause that guides the process towards some good end. I call this the *architectonic model* of generation owing to the regulatory role it assigns to the primary efficient cause in the process. I spend considerable time in Chapter 8 showing how the architectonic model, which is introduced in *GC* II 9 and developed in the biological works, has its roots in the account of generation in Plato's *Timaeus*.

The central argument of the book, which is laid out fully in Chapters 2 to 8, is bookended by two other discussions not mentioned thus far (Chapters 1 and 9). Chapter 1 opens the book with some preliminary puzzles confronting Aristotle's theory of substantial generation. First and foremost is the argument of Parmenides, which challenged the very possibility of substantial generation itself. Aristotle holds that all change occurs between a pair of contraries. In the case of generation and destruction these are being and non-being. Thus, if anything comes to be in the strict sense, it must come to be from non-being in the strict sense (*GC* I 3, 317a32–b5). Parmenides famously denied that this was possible on the grounds that non-being necessarily does not exist. Aristotle confronts this challenge head on in *Physics* I 8 and again in *GC* I 3. In the former, Aristotle responds to Parmenides by conceding that nothing comes to be from what is not in the strict and unqualified sense. But Aristotle insists that things can still come to be from what is not accidentally insofar as they come to be from the privation (e.g. Socrates comes to be healthy from what is not-healthy) (191b13–16). But this response raises a problem. In *GC* I 3 Aristotle says that *if* something comes to be in the unqualified sense, then it must come to be from what is not in the unqualified sense (*to mê on haplôs*) (317a32–b5). Since Aristotle is clear in *GC* that some things do come into being without qualification, namely substances, it follows that substantial generation must proceed from what is not without qualification. Yet that is precisely what *Physics* I 8 appears to deny. One of the main projects of Chapter 1 is to reconcile this tension, since it bears on the very question of whether Aristotle thinks substantial generation is even possible. Like most scholars, I believe that *Physics* I 8 and *GC* I 3 are ultimately compatible. However, while most see Aristotle's two replies to Parmenides as different sides of the same coin, on the reading I defend *GC* I 3 represents an advancement over *Physics* I 8 arising from a development in his thinking about non-being.

I close the book in Chapter 9 by taking a step back and examining the place of substantial generation in Aristotle's broader cosmology. Interpretations of Aristotle on the scope of teleology have tended to gravitate towards one of two opposing extremes. At one extreme is the Cosmic Teleology reading according to which all the parts of the universe, including living things, are coordinated in such a way that their mutual interactions contribute to the good of the cosmos as a whole and, ultimately, the good of man.[31] At the other extreme is what I call the Organism-Centred view according to which all final causation (in biology at least) can be understood exclusively from the perspective of the good of the individual organism itself. On this reading, each individual has the features it does because, and only because, those features are good for that individual.[32] While I count myself among those who accept some version of the latter interpretation, I argue that there is value in considering what Quarantotto (2015) calls 'the view from above' insofar as it forces us to take seriously the connections between Aristotle's biology and his broader cosmology. Chapter 9 is an attempt to accommodate this perspective within the Organism-Centred view by showing that Aristotle's argument at the beginning of *GA* II 1 about the final cause of animal generation is

[31] Sedley (1991, 180; 2010, 24).

[32] Balme (1972[1992], 96–7), Wardy (1993), Bodnar (1997, 2005), Lennox (2001a, e.g. 182–4; 2001b, 341), Judson (2005, 348, 359–60), Gotthelf (2012), Leunissen (2010).

intended to show that *continuous biological generation itself* exists not (or not only) because of the contribution it makes to the individual's own good but because of its contribution to the good state of the universe as a whole.[33] At the same time, I argue that this does not vindicate the Cosmic Teleology interpretation at least in the case of living things. For this is the only place in the corpus where Aristotle adopts a global teleological perspective in order to explain some feature of the living world.

[33] Cf. *Metaph.* XII 10.

CHAPTER I

Preliminary Puzzles

1.1 The Eleatic Challenge

In the *Phaedo* (Plato's) Socrates develops an argument for the immortality of the soul that draws on what has come to be known as the Law of Opposites (70d6–72a7):

P1 Everything that comes to be necessarily comes to be from its opposite.
P2 Between every pair of opposites, there are two processes: from *F* to *G* and from *G* to *F* (e.g. growing is a process from small to large, shrinking from large to small).
P3 Life and death are opposites.
C Therefore, the dead must come to be from the living by dying and the living must come to be from the dead by being reborn.

This argument is meant to provide sufficient (*hikanon*) proof that the souls of the dead continue to exist somewhere whence they come back again (72a6–7). Whether or not Aristotle himself accepts that (some part of) the human soul is immortal (*DA* III 5), he does accept the model of change that underwrites this argument. Aristotle holds that every change involves a subject *S* that undergoes the change and a pair of opposites *F* and *G* that represent the two poles of the change (*Ph.* I 5–7). Since the contraries of generation and destruction are being and non-being (*GA* 741b22–4, *Metaph.* 994a27–8, *Ph.* 224b8–10, 225a15–16), if anything comes to be in the strict sense, it must come into being from non-being. It was this fact that led the Eleatic philosophers to reject the very idea of generation and destruction. For it seems to imply that something can come into being from, and pass away into, absolutely nothing.

The challenge to generation posed by the Eleatics originated in the philosophy of Parmenides. And the starting point of that challenge was the denial of the very possibility of non-being:

> But come now, I will tell you – and you, when you have heard the story, bring it safely away – which are the only routes of inquiry that are for

> thinking the one, that it is and that it is not possible for it not to be, is the path of Persuasion (for it attends upon Truth), the other, that it is not and that it is right that it not be, this indeed I declare to you to be a path entirely unable to be investigated: For neither can you know what is not (for it is not to be accomplished) nor can you declare it. (Fr. 2, Curd translation)

Plato diagnoses the problem raised by Parmenides as, in part, a problem of reference: 'To what should the name, "that which is not" (*to mê on*), refer? . . . What would he [the speaker] indicate by it to someone else who wanted to find out about it?' (*Sophist* 237b10–c4).[1] When we try to say anything about non-being, either we refer to something, in which case what we are referring to is not nothing (237c8), or we refer to nothing, in which case we are not saying anything at all: 'It is absolutely necessary, it seems, that someone who does not say something says nothing at all . . . Therefore, don't we have to refuse to admit that a person like that speaks but says nothing? Don't we have to deny that anyone who tries to utter "what is not" is even speaking?' (*Sophist* 237e1–6).[2] Furthermore, Plato argues that even if one could somehow refer to non-being, it could not function as a proper subject of predication, and so we could not form any meaningful sentences about it (238a1–c10).[3]

Plato responds to Parmenides by abandoning the sense of non-being that means *what does not exist* (the existential sense) and replacing it with the concept of *difference*:

> ELEATIC STRANGER: It seems that when we say 'that which is not' we don't say something contrary (*enantion*) to 'that which is' (*tou ontos*) but only something different from it.
>
> THEAETETUS: Why?
>
> ES: It's like this. When we speak of something as not large, does it seem to you that we indicate the small rather than the equal?
>
> THT: Of course not.
>
> ES: So we won't agree with someone who says that negation signifies a contrary. We'll only admit this much: when 'not' and 'non-' are prefixed to names that follow them they indicate something *other* than the names, or rather, something *other* than the things to which the names following the negation are applied.
>
> (*Sophist* 257b3–c3, White translation)

1 For discussion of Plato on non-being in the *Sophist* see Ferejohn (1989), Brown (1986 [1999], 1994), Owen (1999), Notomi (1999, 173–9).

2 Plato puts this same argument in the mouth of the sophists, Dionysodorus and Euthydemus, at *Euthydemus* 283c–e, which may tell us something about his attitude towards that puzzle there.

3 Compare *Theaetetus* 188e–189b.

By the contrary of being, the Stranger means absolute non-existence (what Aristotle calls 'unqualified non-being', *to mê on haplôs*, 'what completely does not exist', *to mê on holôs*, or 'the universal negation of everything', *apophasis estai katholou pantôn: GC* 317b7–13). He agrees with Parmenides that there is no such thing as *to mê on* in that sense; he grants Parmenides at least this much when he says 'good-bye' to the contrary of being:

> Nobody can say that that which is not, which we've made to appear and now dare to say exists, is the contrary of that which is. We've said good-bye long ago to any contrary to being, and to whether it is or not, and also to whether or not an account can be given of it. (258e7–259a1)

Rather than treat non-being as nothing (as no existing thing), the Stranger instead proposes they take it to refer to the class of existing things that are merely *different* from, or *other* than (*heteron*), the thing that is negated by the expression (257e2–258c4). In other words, the phrase "what is not" is now taken to mean "what is not-*F*", where this refers to all of those things that are not the *F* in question.[4] For example, the not-beautiful (*to mê kalon*) refers to any existing thing that is different from beauty and the not-large (*to mê mega*) to any existing thing that is different from largeness (257e7–258a9). I shall argue that this same strategy lies behind Aristotle's own response to Parmenides in *Physics* I 8 (see Section 1.2).

One of the main consequences of Parmenides' repudiation of the concept of non-being was the denial that anything comes to be or passes away. Aristotle reconstructs the basic Eleatic argument as a two-horned dilemma (*Ph.* I 8, 191a23–31).[5] If what is came to be, then it must have come to be (1) either from what is or (2) from what is not (since these are contradictory opposites: 191b26–9). (1) It could not have come to be from what is, since it was already there. (2) Nor could it have come to be from what is not, because something must pre-exist the change.[6] Call this the Eleatic

[4] This is the so-called Oxford interpretation of Plato on negative predication. For a more restrictive reading of the scope of the *F* in "what is not-*F*" see Ferejohn (1989).

[5] While the first horn of the dilemma is explicit in Parmenides' Fr. 8, the second horn is not. But Aristotle does not present the dilemma as a faithful reconstruction of Parmenides' own reasoning. Instead, I agree with Anagnostopoulos (2013 n. 1) that he thinks he is simply expressing the broader Eleatic position.

[6] The formulation of the puzzle in *GC* I 3 makes it clear that *hupokeisthai* at 191a31 refers to the need for a *pre-existing* subject (cf. *GC* 317b15–18: *prouparchein*; 317b29–31: *prouparchontos*) not the need for something to persist as an 'underlier' (Charles 2018, also Anagnostopoulos 2013, 262). Loux objects to reading *hupokeisthai* here as 'pre-existing' on the grounds that it would not provide a separate reason for endorsing the claim that nothing comes to be from what is not but 'merely reformulates the claim or restates it in other words' (1992, 285). I disagree. So understood, Aristotle is saying (a) that there

Challenge. Before we can investigate the nature of generation, we first need to show that generation is possible. And that means resolving this dilemma. At *Physics* I 8, 191b27–9 Aristotle identifies two ways of resolving it. *Physics* I 8 itself presents one way of solving the puzzle, which (I shall argue) has its roots in the account from Plato's *Sophist*. The other way is generally thought to come in *GC* I 3, which makes use of the distinction between potentiality and actuality. I shall take up these two accounts in turn. While most scholars treat these two ways of resolving the dilemma as two versions of the same solution presented from two different perspectives (e.g. Algra 2004), I shall argue that they are independent solutions. Yet, it is not obvious that these two solutions are entirely compatible.[7]

The analysis that follows is difficult and at times quite abstract, since it deals with complex metaphysical issues surrounding the concepts of being and non-being. Those unfamiliar with the relevant debates may quickly find themselves 'laid low like seasick passengers', as Socrates says in the *Theaetetus* (191a4–8). Yet, it is necessary that we work our way through the puzzles in order to clear the way for Aristotle's positive account of substantial generation in the rest of this book. For Aristotle admits that Parmenides' challenge makes it 'extraordinarily difficult to see how something could come to be without qualification' so that one might 'raise a puzzle' (*aporêseie*) about whether substances come into being at all (*GC* I 3, 317b23–33). While I try my best to make this dense material accessible, those who start to feel queasy may wish to skip ahead to Chapter 2.[8]

must have been something there before the emergence of being and (b) that this could not have been absolute non-being since that kind of non-being necessarily does not exist. These are clearly not the same claim.

[7] Compare Kelsey (2006, 339), who takes the two solutions to be independent but consistent accounts. I fall somewhere closer to him on this issue.

[8] While this chapter is aimed at those who are not familiar with the relevant debates, for those who do skip ahead there are two main takeaways from this chapter that contribute to our understanding of substantial generation going forward. First, the account of non-being in *GC* I 3 (Section 1.3) tells us something about what matter must be like in order to serve as a subject of generation, namely, it must be a substance 'in potentiality' (*dunamei*) but not a substance 'in a state of completion' (*entelecheia*). I take this to mean that generation proceeds from a pre-existing subject that 'is not' in the sense that it is not actually *the substance at the end of the change* (e.g. not a human or a horse), though it 'is' that kind of substance in potentiality. Second, Aristotle tells us that this subject must be a determinate individual (a 'this something', *tode ti*) that is capable of separate existence (cf. *GC* 317b7–11, b23–34). I argue that this further requirement rules out the possibility that Aristotle has anything like the traditional concept of prime matter in mind as the underlying subject of generation. Instead, the subject from which substantial generation proceeds must be some sort of proximate matter (e.g. fertilised menstrual blood).

1.2 Aristotle's First Response: *Physics* I 8

Aristotle first introduces the Eleatic Challenge at the outset of *Physics* I 8, which he formulates as a two-horned dilemma:

> The first of those who studied philosophy were misled in their search for truth and the nature of things by their inexperience, which, as it were, thrust them into another path. So, they say that none of the things that are either comes to be or passes out of existence, because what comes to be must do so either from what is or from what is not, both of which are impossible. For (1) neither can what is come to be <from what is> since it already exists, (2) nor can it come to be from what is not since something must exist beforehand. (191a23–31, Hardie and Gaye translation with modifications)

As noted, Aristotle presents the Eleatic Challenge as a two-horned dilemma:

(1) What is could not have come to be from what is, because it was already there.
(2) What is could not have come to be from what is not, because something must pre-exist the change.

When it comes to interpretations of *Physics* I 8 the main axis of division revolves around Aristotle's use of the verb "to be". By "what is" and "what is not" does Aristotle mean "what exists" and "what does not exist" or "what is-*F*" and "what is not-*F*"?[9] Of course there is no reason to suppose that Aristotle's use of the verb must be uniform throughout the chapter. In what follows I defend a hybrid reading. On this reading, Aristotle presents the dilemma in terms of what exists/what does not exist but his solution to it appeals to what is *F*/what is not-*F*.[10] Let me begin with the first part of this.

[9] See Introduction (note 11). For convenience I shall assume that the complete/existential and incomplete/predicative uses of 'is' are both dichotomous and exhaustive. While this is not universally agreed (for a summary of the issues see Brown 1994), the nuances of that debate are not relevant to this chapter.

[10] This is how I read the *Sophist*. The *Physics* account admits of a slightly more nuanced version of this hybrid reading, since Aristotle's solution to the second horn of the dilemma (191b13–16) can itself be broken down into two claims (cf. Anagnostopoulos 2013, 261–2): (i) it is not possible for something to come to be from what is not in the unqualified sense (b13–14); (ii) it is possible for it to come to be from what is not in the accidental sense (b14–15). A purely existentialist reading would take "what is not" in both (i) and (ii) to mean what does not *exist*, while a purely predicative reading would take them both to mean what is *not-F*. On the more nuanced hybrid reading, "what is not" means what does not exist in (i) and what is not-*F* in (ii). While I favour this more nuanced reading of Aristotle's solution, my interpretation is compatible with the purely predicative reading. I agree with Anagnostopoulos (2013, 261) that a purely existential reading of *Physics* I 8 does not make sense.

I follow the orthodox view that by "what is" and "what is not" Parmenides meant what *exists* and what *does not exist*.[11] If this is right, then it is reasonable to suppose that Aristotle too understood the original Eleatic puzzle in those same terms. This is almost certainly how Plato understood Parmenides (see above). In that case we can take the second horn of the dilemma to be a straightforward rejection of generation *ex nihilo* (something cannot come into being from absolutely nothing, cf. *GC* 317b29–31). Consider the Big Bang, which is when the known universe first began. What existed before the Big Bang? We share Parmenides' intuition that *something* must have pre-existed the Big Bang; the universe could not have just popped into existence from absolutely nothing. The first horn of the dilemma is trickier. However, the argument makes good sense if we again take "what is" to mean what exists generally (cf. *GC* 317b7: 'what is universal and includes everything').[12] What exists could not have come into existence from some pre-existing thing, since the whole idea that being *came to be* presupposes that it did not already exist. If something existed prior to the Big Bang, then it does not make sense to describe that event as the moment when being came into existence; for it was already there beforehand.[13] In the remainder of this chapter I will focus exclusively on the second horn of the dilemma that what is cannot come to be from what is not, since that appears to be the horn that most exercised Greek philosophers after Parmenides. Aristotle tells us in *GC* I 3 that the possibility that coming-to-be might proceed 'from no pre-existing thing' (*to ek mêdenos gignesthai prouparchontos*) was the one consequence of the Eleatic position that 'more than any other preoccupied and alarmed the earliest philosophers' (*GC* 317b29–31).

According to the second horn of the dilemma, what is cannot come to be from what does not exist since there must be some pre-existing subject from which the change proceeds (as Aristotle has argued in *Physics* I 7).[14] The heart of Aristotle's reply to this horn comes at 191b13–16:

[11] Brown (1994, 216–20).

[12] Cf. Williams (1982, 82): '"What is universal and includes everything" is "being" in the most general sense: whatever can in any sense be said to be, "is" in this sense.'

[13] Aristotle is not especially concerned with the "Big Bang" problem, since his universe is eternal. According to his cosmology, everything that exists has always existed and will continue to exist in the future, and so there was never a time for which we would have to ask what preceded being. However, *Physics* I 8 shows that Aristotle still thinks we need to address the Eleatic Challenge. As we shall see, his own solution(s) does not depend on his view that being is eternal.

[14] Parmenides may have held this on the grounds that something must have generated what exists. But *Physics* I 7–8 does not rely on that assumption. Instead the emphasis is on the need for a pre-existing subject (*to hupokeimenon*). By "subject" here Aristotle simply means that *from which* the change originally proceeds non-accidentally (e.g. a seed). See Chapter 2.

> We ourselves say that nothing comes to be from what is not in the unqualified sense, though it is surely possible to come to be from what is not in a way, namely, in an accidental sense. For it comes to be from the privation which in itself is not, this not surviving as a constituent in the thing that comes to be.

Aristotle's solution here turns on the technical distinction he drew back at 191a35–b9 between two ways in which one thing comes to be "from" another, namely, in virtue of itself (Greek *kath' hauto*; Latin *per se*) versus incidentally/accidentally (Greek *kata sumbebêkos*, Latin *per accidens*).[15] Suppose Hippocrates (a doctor) is also trained as an architect so that the art of building and the art of medicine just happen to coincide in one and the same individual. When Hippocrates builds a house, there is a sense in which a house comes to be from a doctor, not in virtue of being a doctor, but incidentally insofar as the doctor also happens to be a house-builder. Building houses is incidental to the function of a doctor (*Metaphysics* VI 2, 1027a1–4). Put formally, we say that A comes to be from C incidentally if A comes to be from B in virtue of itself but B also happens to be C (where being C is incidental to B). Now consider the case in which healthy-Socrates comes to be from unhealthy-Socrates, which is a combination of a subject and a privation. Since *Physics* I 8 treats the privation as a kind of non-being (e.g. illness is non-health), and since this privation is incidental to Socrates, it follows that *in a way* what is comes to be from what is not, not in virtue of itself, but only accidentally. For Aristotle holds that a thing comes to be *per se* from the subject but only *per accidens* from the privation insofar as the privation is an accident of the subject (cf. *Ph.* I 5, 188a32–b2). It is this part of the doctrine of *Physics* I 7 that Aristotle means to draw on in I 8 in order to resolve the Eleatics' puzzle (cf. 191a24–5):

> Clearly, then, if there are causes and principles of naturally existing things from which (*ex hôn*) they primarily are and have come to be (I mean what each is said to be according to its essence and not incidentally), then everything comes to be from both the subject and the form ... Now the subject is one in number but two in form. For there is the man, the gold, and in general the countable matter; for it [the subject] is more of the nature of an individual "this", and what comes to be does not come to be from it incidentally, while the privation and the contrariety merely coincide (*sumbebêkos*) with it. (190b16–27)

[15] With the following see *Metaphysics* V 7 and VI 2. The distinction is well-rehearsed in the literature (e.g. Loux 1992, Meyer 1992).

On my interpretation, *Physics* I 8 can be understood as employing the same general strategy that Plato adopts in the *Sophist* (see above). Both concede to Parmenides that nothing comes to be from what is not without qualification. But both hold that things can come to be from what is not in some qualified way insofar as they come to be from what is not-*F* (e.g. what is not-healthy or not-beautiful). In this way both accounts seek to avoid the Eleatic Challenge by replacing the complete and existential use of "is (not)" contained in its original formulation with its incomplete and predicative use.[16] Aristotle's contribution to this debate lies in applying his own intrinsic/accidental distinction to that analysis in order to differentiate the ways in which a thing can and cannot come to be "from" what is not. In Section 1.3 I will show how this the strategy is quite different from the one adopted in *GC* I 3, where Aristotle appears to concede much less ground to the Eleatics.

1.3 Aristotle's Second Response: *Generation and Corruption* I 3

At the end of *Physics* I 8 Aristotle identifies 'another way' (*allos d'*) of resolving the dilemma that turns on the distinction between potentiality and actuality (191b27–9). Most commentators assume the reference is to *GC* I 3. The main focus of *GC* I 1–4 is the question of whether there is such a thing as substantial generation or whether all such cases can be reduced to a form of qualified change.[17] It is in the context of answering this question that Aristotle again raises the puzzle about non-being in *GC* I 3:

> [A] Having distinguished these things, we must first consider whether there is anything that comes to be and passes away in the unqualified sense or whether nothing comes to be in the strict sense (*kuriôs*) but everything always comes to be something from being something (e.g. comes to be healthy from being sick and sick from being healthy or comes to be small from being big and big from being small, and so on in every other instance). [B] For if there is coming-to-be without qualification, then something must come to be from what is not without qualification so that it would be true to say that not-being belongs to some things. [C] For coming-to-be in the qualified sense proceeds from what is not in the qualified sense (*mê ontos tinos*), e.g. from what is not white or not beautiful, while coming-to-be in the unqualified sense proceeds from what is not in the unqualified sense (*haplôs mê ontos*). (*GC* I 3, 317a32–b5)

[16] One difference is that (at least on the so-called Oxford interpretation) in the *Sophist* Plato takes "what is not-*F*" to be satisfied by any existing thing that is different from *F* (e.g. largeness counts as "what is not-beautiful") whereas Aristotle restricts it to the specific privation of *F* (e.g. ugliness). Compare *Physics* I 5–6.

[17] See Chapter 3.

Aristotle is explicit here that if anything comes to be in the strict and unqualified sense, then it must come to be from unqualified non-being: 'for coming-to-be in the qualified sense proceeds from what is not in the qualified sense, e.g. from what is not white or not beautiful, while coming-to-be in the unqualified sense proceeds from what is not in the unqualified sense.' Aristotle's examples of qualified non-being here make it clear that he has in mind the privation. If this is right, then *GC* I 3 means to contrast cases where a thing comes to be from the privation (qualified becoming) with those cases where it comes to be from unqualified non-being (coming-to-be in the strict sense, i.e. substantial generation).[18]

We can already see how *GC* I 3 departs from *Physics* I 8. In *Physics* I 8 Aristotle denies that anything comes to be from what is not in the unqualified sense. But he does allow that things can come into being from what is not in some qualified and accidental way insofar as they come to be from the privation (191b13–16). While Aristotle continues to associate the privation with a kind of qualified non-being in *GC* I 3, he now restricts that to cases where a thing comes to be in the qualified sense (*genesis tis*). Substantial generation is then marked off from this as a distinct form of change that proceeds from unqualified non-being – something that Aristotle explicitly denied was possible in *Physics* I 8. *GC* I 3 thus faces the second horn of the Eleatic Challenge head on by showing that some things do come to be from what is not in the unqualified sense, namely, substances. The way Aristotle does that, I will argue, is by attempting to rescue the concept of unqualified non-being from the Eleatics in a way that avoids their worries about generation *ex nihilo*.

Let me start again with *GC* I 3, 317a32–b5. The main point of that passage is that, if there is such a thing as unqualified generation – and Aristotle clearly thinks there is – then it must proceed from unqualified non-being (317a30–b5). But what does "unqualified non-being" (or "what is not without qualification") refer to in this context? Aristotle goes on to say that "unqualified" may be taken

[18] One might object here that in *Physics* I the concept of a privation is meant to cover both kinds of non-being in [C]. In that case Aristotle continues to hold that in every case of coming-to-be a thing comes to be from the privation, as the *Physics* I 8 solution suggests: qualified coming-to-be proceeds from the privation of some non-substantial property (e.g. the lack of health), which counts as qualified non-being; unqualified coming-to-be proceeds from a privation of the substantial form, which counts as unqualified non-being. However, in *Physics* I 7 Aristotle characterises the privation as *an accident* of the subject (190b26–7). And so it is unlikely that he means to include under that concept the privation of a substantial form; for that type of lack is not an accident of the subject.

to signify either: (i) 'what is primary among [in?] each category of being';[19] or (ii) 'what is universal and includes everything' (317b5–7). Taken in the first way it refers to what is not *a substance* (the primary category of being), which at this stage in the argument is meant to signify something in the non-substantial categories (e.g. a quality or quantity). Aristotle argues that substantial generation cannot proceed from what is not in this sense since that would entail that non-substantial items could exist in their own right in separation from substance (317b7–11).[20] Suppose the first thing to come into being was a red apple. Aristotle's point is that the apple (the substance) could not have come to be from the redness (a non-substance) – in the language of *Physics* I 7, the redness cannot serve as the pre-existing subject from which the change proceeds – because a quality cannot exist in its own right in separation from a substance; likewise for all non-substantial items. On the second reading 'unqualified non-being' refers to what completely does not exist or the universal negation of all being, i.e. absolute non-existence (317b11–13).[21] This option is also ruled out as the *terminus a quo* of generation on the grounds that it would imply that something could come into being from absolutely nothing at all. And Aristotle agrees with the Eleatics that there is no such thing as generation *ex nihilo*. Thus, Aristotle argues that if we take unqualified non-being (the subject of unqualified generation) in either of these two ways, then it is difficult to see how substances could come into being at all (cf. 317b21–2). For he says that *if* anything comes to be without qualification, then *it must* come to be from what is not without qualification.

The upshot of this (admittedly difficult) analysis is that substantial generation must proceed from something that is itself a substance (since non-substantial items cannot serve as pre-existing subjects of change), but one that lacks being in some yet-to-be-determined sense (since substantial generation proceeds from unqualified non-being), yet does not lack it in a sense that would render it completely non-existent (since something cannot come into being from nothing).

To resolve the puzzle Aristotle proposes that we think of substantial generation as proceeding in one way from what is not though in another way from what is, so that the subject of generation can be spoken of both as a being and as a non-being. It is at this point that he appeals to the

[19] For two different interpretations of *to prôton kath' hekastên katêgorian tou ontos* see Williams (1982, 82). My reading is compatible with both of those.

[20] It is standard Aristotelian doctrine that only (primary) substances are capable of separate existence (existing in their own right) (*Categories* 5, cf. *APo.* I 4).

[21] See Philoponus, *in GC* 46,30–47,3.

distinction between potentiality and actuality, which is supposed to help disarm the (second horn of the) Eleatic Challenge:

> In one way things come to be from what is not in the unqualified sense, yet in another way they always come to be from what is. For there must be something existing beforehand (*prouparchein*) that potentially *is* but actually *is not*, and this is spoken of both as being and as not-being. (317b15–18)

He repeats this a few lines later:

> For if something comes to be [sc. in the unqualified sense], then it is clear that there will be a certain substance in potentiality (*estai dunamei tis ousia*) but not in actuality (*entelecheia d' ou*), from which generation will necessarily proceed and into which what is destroyed is necessarily transformed. (317b23–25)

According to this new analysis, unqualified non-being refers to something that *is* a substance in potentiality but *is not* a substance in actuality. This allows Aristotle to hold that substantial generation proceeds from what is not in the unqualified sense while avoiding the consequences arising from the other two senses of 'unqualified' non-being above. It avoids the problem of generation *ex nihilo*, since there is some being that pre-exists the change. It also allows Aristotle to accept that a substance comes to be from what is not a substance – the first sense of 'unqualified' non-being (317b5) – without committing himself to the idea that properties can exist in separation from substances. For substantial generation proceeds from what is not a substance in actuality, though it is a substance in potentiality.

The *GC* I 3 account thus countenances three kinds of non-being:

(1) the privation (what is not-*F*)
(2) absolute non-existence (what does not exist)
(3) potential being (what is not a substance in actuality but is a substance in potentiality).

As we have seen, Aristotle treats (1) as a form of qualified non-being (*to mê on ti*), while (2) and (3) count as non-being in the unqualified sense (*to mê on haplôs*). However, only (3) can serve as the proper subject of coming-to-be without qualification. Thus, in one sense substantial generation proceeds from what is not in the unqualified sense but, in another way, it always proceeds from what is: 'For there must be something existing beforehand that potentially *is* but actually *is not*, and this is spoken of both as being and as not-being' (317b15–18).

Even with this distinction in hand Aristotle says that 'it is extraordinarily difficult to see how a thing can come into being in the unqualified sense' (317b18–20). For it is unclear how we ought to characterise this potential being that is at the same time not any actual being. And so we still cannot say for certain whether or not substances come into being at all (b21–2). We thus need to say something about the nature of this potential being that serves as the subject from which substantial generation proceeds.

1.4 Potential Being

According to one popular interpretation, Aristotle holds that the matter for substantial generation is something called prime matter.[22] This is not ordinary matter; it is a metaphysical postulate that lacks any determinate properties of its own but exists as pure potentiality (understood purely as a capacity to become all things). At least as I read *GC* I 3, this is not Aristotle's answer. For while he considers something like prime matter as a possible candidate for unqualified non-being, he rejects it on the grounds that it leads to a 'remarkable puzzle' (*thaumastên aporian*, 317b18–19) whose consequences are unacceptable:

> [1] For if something comes to be, then it is clear that there will be a certain substance in potentiality but not in actuality, from which generation necessarily proceeds and into which what is destroyed is necessarily transformed. [2] Then will any attribute belonging to the remaining categories belong to this in actuality? I mean, for example, will that which is only potentially a determinate individual (*tode ti*) and a being but not one without qualification possess any determinate size or quality or position? For if it possesses not even one of these <in actuality> but all of them potentially, then [2.1] what is not in this sense turns out to be separable (*chôreston*), and further [2.2] generation proceeds from no pre-existing thing (*to ek mêdenos gignesthai prouparchontos*), a consequence which, more than any other, preoccupied and alarmed the earliest philosophers. [3] If, on the other hand, it possesses some of the remaining properties mentioned earlier but these do not belong to some determinate individual or substance (*tode ti ê ousian*), then properties will exist in separation from substances. (*GC* I 3, 317b23–33)

This text is difficult to unpack. Section [1] begins by stating Aristotle's preferred analysis of unqualified non-being as something that *is not* a substance in actuality but *is* a substance in potentiality. Aristotle then

[22] See Joachim (1926, 97, 104).

goes on to reject two ways of conceptualising this non-being, both of which come close to the traditional idea of prime matter:

A Unqualified non-being refers to a kind of pure potentiality: a *potential* substance that has no actual properties (e.g. no actual size or quality or position) but has them all potentially.

B Unqualified non-being refers to a potential substance that is nevertheless endowed with actual properties.

Section [2] targets A while section [3] targets B.

In [2.2] Aristotle argues that A commits us to the view that substances come into being from something that is potentially everything but actually nothing, which he sees as a no-less-noxious version of generation *ex nihilo*.[23] Arguments [2.1] and [3], on the other hand, both turn on the notion of separability. By calling something 'separate' Aristotle means it is the sort of thing that is capable of existing in its own right (*kath' hauto*) as opposed to those things that are only capable of existing in virtue of being predicated of some further underlying subject (*kath' hupokeimenou*) (*APo.* I 4, 73b5–10). Colours, for example, cannot exist separately but only by inhering in some underlying substance (e.g. an apple), whereas individual substances can exist separately (in their own right) without having to inhere in something else.[24] Text [2.1] rejects the possibility of a purely potential substance standing as the *terminus a quo* of generation because it implies that what is actually nothing at all could exist in its own right, while text [3] rejects the idea of a potential substance endowed with actual properties because it implies that properties could exist on their own independently of any actual substance underlying them (cf. 317b7–13). But properties, we have seen, can exist only by being predicated of some actual substance.

This appears to rule out the concept of prime matter as the subject of generation.[25] While prime matter counts as a potential substance, it is incapable of separate existence and so falls foul of [2.1]. Even if we concede that prime matter never exists in its own right but is always endowed with particular properties (Joachim 1926, 97), Aristotle rules that out in [3] on the grounds that what is only potentially a substance cannot serve as an anchor for properties. That conception of prime matter entails that

[23] Philoponus, *in GC* 48,24–6.

[24] In the *Metaphysics* the concepts of separability, independent existence, and individuality (being a "this") are all interconnected (e.g. 1028a23–4, 1028a33–4, 1029a28). See Philoponus, *in GC* 30,24–29.

[25] For a similar argument see Gill (1989, 42–6).

properties could exist independently without being tethered to any determinate substratum.[26] Thus, while prime matter counts as 'a certain substance in potentiality but not in actuality', it cannot be the sort of non-being that is required to serve as the *terminus a quo* of substantial generation. Instead, *GC* 317b23–33 implies that generation must proceed from some actual pre-existing substance with actual properties of its own, since this alone is capable of existing in its own right.

Fortunately, there is another way of conceptualising unqualified non-being that is available to Aristotle that meets the conditions set out above (it must be an actual substance that lacks being but not in the sense that would render it completely non-existent). On this alternative *GC* 317b23–33 is implicitly defending the view that the potential being from which generation proceeds is some kind of proximate (rather than prime) matter.[27] For example, biological generation begins from a seed which is a determinate substance with actual properties of its own and so enjoys the kind of independent existence that Aristotle demands of the subject of generation: it exists in its own right (*kath' hauto*) and not merely as something predicated of a substance.[28] It also lacks being in the relevant sense. In text [1] Aristotle says that the non-being from which substantial generation proceeds is not a substance *entelecheia* but *dunamei* (317b23–33). It is well known that Aristotle coined the term *entelecheia* to express what is at root a teleological concept. The idea has to do with being complete or perfect (*teleion*) and is associated with the form of a thing when it has achieved its fully developed state (*Metaph.* V 16, 1021b22–30).[29] Thus, in the case of things that come into being, we say that a thing exists *entelecheia*

26 Back at *GC* 317b7–13 Aristotle gave a similar argument to [3]: 'But what is not a substance or a "this" clearly cannot possess items from any of the other categories (e.g. neither quantity nor quality nor position); for properties cannot exist in separation from substances.' The argument in [3] implies the stronger claim that what is only *potentially* but not *actually* a substance cannot possess attributes either. Otherwise the conclusion he draws in [3] would not follow. In that scenario properties would not exist separately but would be predicated of that potential substance. So [3] makes it clear that Aristotle thinks the idea of a potential substance that possessed actual properties of its own is incoherent. For a similar point see Gill (1989, 44).

27 This reading is defended by Gill (1989, 42–4) and Algra (2004). Although Algra ultimately rejects the concept of prime matter in favour of proximate matter, he argues that the two options are not necessarily incompatible (2004, 105). But the idea that the *terminus a quo* of generation might be a potential substance that is already informed is exactly what Aristotle rejects in [3].

28 A seed will not count as a substance in the full sense of the concept developed in the central books of the *Metaphysics*. But I have already argued in the Introduction that in *GC* Aristotle is working with the less technical concept of substance as an individual "this" (or *tode ti*) that is capable of existing in its own right. A seed is a substance in that sense.

29 On the significance of the Aristotelian technical term *entelecheia* see Johnson (2005, 82–90), Beere (2009, 218–19). The link between possessing a substantial form and existing in a state of completion is critical to Aristotle's conception of soul in *De anima*.

when it is in a state of completion, where the completion in question is defined by the form that constitutes the end for the sake of which it came to be. For example, a house exists in a state of completion when it is capable of providing shelter. At that point we say that it has attained its good and is a perfect specimen of its kind (cf. *Metaph.* V 26, 1021b12–14). This suggests that when Aristotle says in text [1] that substantial generation proceeds from something that *is* a certain substance in potentiality but *is not* a substance in a state of completion, he means what is not yet the substance towards which that change is directed. For example, the generation of an oak tree proceeds from a seed that *is* an oak tree in capacity or in potentiality but *is not* an oak tree in a state of completion. For it does not yet possess the kind of substantial being that makes a living thing an oak tree.[30]

In this way *GC* I 3 answers the second horn of the Eleatic Challenge (what is cannot come to be from what is not because something must pre-exist the change), not by conceding that things only come into being from what is not in some *qualified* and *accidental* way (insofar as they come to be from the privation), but by reconceptualising the concept of *unqualified* non-being in terms of a kind of *incomplete* being.[31]

1.5 Are the Two Solutions Compatible?

It should be clear by now that there is a significant gap between *Physics* I 8 and *GC* I 3 over the question of whether it is possible for anything to come into being from what is not without qualification. From the perspective of *GC*, at least, this affects the very possibility of substantial generation. For Aristotle says there that *if* something comes to be without qualification, then it must proceed from unqualified non-being (317a32–b5: section [B]) and that, further, if we cannot make sense of this, then the very notion of substantial being is called into question (317b18–22). This conflict is not something that modern commentators have taken seriously enough.[32] Put simply: *Physics* I 8 concedes to the Eleatics that nothing comes to be from

[30] At *GA* 729a32–3 Aristotle tells us that menstrual blood can be characterised as *prôton hulê*. I take this to mean that it is the *first* matter used to construct the original embryo, as opposed to the subsequent matter that is used to grow the parts later on (*GA* 744b32–745a4). For a slightly different reading see Connell (2016, 153). In Henry (2006b n. 3) I had taken *prôton hulê* as a reference to so-called prime matter, though I have long since abandoned that view.

[31] I use "incomplete" here to contrast the teleological notion of existing in a state of completion (*entelecheia*). I do not mean the incomplete use of the verb "to be" (Aristotle is clear that in this context "what is not" is being used in its complete and unqualified sense).

[32] This problem was first recognised by Williams (1982, 80–1).

what is not in the unqualified sense (191b13–16); instead everything comes to be from what is not only accidentally insofar as it comes to be from the privation (what is not-*F*). By contrast, Aristotle is explicit in *Generation and Corruption* that some things *do* come to be in the unqualified sense and that they do so from what is not without qualification (*to mê on haplôs*), which he explicitly contrasts with those cases where a thing comes to be from what is not only in the sense of the privation (qualified non-being). So it looks like *Physics* I 8 denies, while *GC* I 3 accepts, that unqualified generation is possible.[33]

Algra (2004, 116–20) attempts to explain away the tension by suggesting that the two accounts treat unqualified non-being at 'different levels of analysis'. He argues that *Physics* I 8 treats the concept from an ontological perspective (what the concept is actually meant to pick out in reality), while *GC* I 3 is working within the framework of ordinary language and is only dealing with what Greek speakers would 'be prepared to call' unqualified non-being. But Aristotle does not present the discussion of unqualified non-being in *GC* I 3 as an analysis of ordinary language, nor does he frame the discussion in terms of reconciling commonly held beliefs (*endoxa*). Instead *GC* 317b7–13 is clearly meant to offer an ontological analysis of the different kinds of non-being marked off by the two senses of the word *haplôs*. Even if we accept that ordinary Greek speakers would have understood *to mê on haplôs* to mean "nothing", as indicated by the second sense of *haplôs*, the first sense ("a non-substance") clearly depends on Aristotle's own technical doctrine of the *Categories*, as do his reasons for ruling out that meaning (317b1–13). While Aristotle does go on to discuss what ordinary people mean by unqualified non-being (318b18–28), that forms no part of his own analysis of substantial generation in *GC*.

[33] The *Physics* I 8 claim also creates a tension within the *Physics* itself. *GC* I 3 holds that if there is coming-to-be without qualification, then something must come to be from what is not without qualification (317b1–5). Since *Physics* I 8 denies that anything comes to be from what is not without qualification, this would imply that nothing comes to be in the unqualified sense. But how do we reconcile that with the claim in *Physics* I 7 that substances alone come to be without qualification (190a32–3)? Kelsey suggests that by the end of *Physics* I 7 Aristotle 'has not yet made it clear whether substances really do come to be' (Kelsey 2006, 343–4 n. 24). In that case we might read *Physics* I 8 as conceding that what passes for substantial generation is in fact reducible to a form of qualified becoming (which conflicts with Aristotle's position in *GC*). Alternatively, Aristotle may simply reject in *Physics* I 8 the principle endorsed at *GC* 317b1–5 that if something comes to be in the unqualified sense it must proceed from unqualified non-being. Instead, he may be thinking that even in cases of substantial generation what comes into being comes to be from what is not only in the accidental sense of coming to be from the privation. At the very least, given the incongruity between *Physics* I 8 and *GC* I 3, I suspect that what *GC* 317b13–15 refers back to is not *Physics* I 8 as most scholars assume (e.g. Algra 2004, 110–15, Burnyeat 2004 n. 25). It may instead be a reference to *Physics* V 1.

Indeed, he dismisses outright this commonsense way of understanding non-being as false (b26–7).

Like most commentators I think that *Physics* I 8 and *GC* I 3 are ultimately compatible. What my interpretation offers is a new way of thinking about how the two accounts fit together. I agree with Algra (2004, 113) that at a very general level the two approaches 'serve to solve the problem of not-being by redescribing the not-being at issue'. However, I have argued that while *Physics* I 8 attempts to redescribe non-being in terms of the privation (a sort of qualified non-being), *GC* I 3 seeks to rescue the concept of unqualified non-being itself by reconceptualising it in terms of what I have called incomplete being. Thus, where Algra sees the two accounts as two sides of the same coin, on my reading *GC* I 3 represents a further development in Aristotle's thinking about non-being. To see this, we need to go back to *Physics* I 8 and ask why Aristotle concedes there that nothing can come into being from what is not in the unqualified sense.

Recall the key text from *Physics* I 8:

> We ourselves say that (A) nothing comes to be from what is not in the unqualified sense, though (B) it is surely possible to come to be from what is not in a way, namely, in an accidental sense. (191b13–16)

Aristotle doesn't say why he endorses (A) in *Physics* I 8. I want to suggest that the analysis of the senses of *haplôs* in *GC* I 3 provides us with a clue. Aristotle first tells us that coming-to-be in the unqualified sense must proceed from what is not in the unqualified sense. He then says that "unqualified" can refer either to some non-substantial item or to absolute non-existence. Nothing can come to be from what is not in the first sense because it implies that properties can exist in separation from substances (317b7–11), while the second is ruled out because it implies that something can come into being from absolutely nothing at all (b11–13) which violates the principle that something must always pre-exist the change (cf. 317b15–18: *prouparchein*; b29–31: *prouparchontos; Physics* 191a31). If Aristotle's understanding of unqualified non-being in (A) is limited to just these two senses, then we can see why he would endorse that claim. For *GC* I 3 tells us that neither of these two senses of non-being can serve as the proper subject from which generation proceeds.

On this interpretation there is no irresolvable conflict between *Physics* I 8 and *GC* I 3 with regard to unqualified becoming. *Physics* I 8 has in mind the kind of unqualified non-being that we find in *GC* I 3 at the stage in the argument where Aristotle gives his *preliminary* analysis of the concept. And

GC I 3 agrees that nothing can come to be from what is not in either of those two senses. But *GC* I 3 goes beyond *Physics* I 8 in offering a third way of conceptualising unqualified non-being, namely, what *is* a substance in potentiality but *is not* a substance in a state of completion. By reconceptualising unqualified non-being in this way, the *GC* I 3 account of generation represents an innovation over the old Platonic way of dealing with the Eleatic Challenge found in *Physics* I 8.

This way of thinking about the relation between *Physics* I 8 and *GC* I 3 on the concept of non-being can be usefully compared to Plato's different analyses of false judgement in the *Theaetetus* and the *Sophist*. In the *Theaetetus* Plato denies that false judgement can be analysed in terms of non-being. One reason may be that in the *Theaetetus* Socrates is focused on absolute non-being (or what is not 'itself by itself'), which he rejects as incoherent (188c12–189b8). It is not until the *Sophist* reconceptualises non-being in terms of difference or otherness that we finally get Plato's solution to the puzzle. Likewise, *Physics* I 8 rejects the possibility of coming-to-be from what is not in the unqualified sense. The reason, I have suggested, is that there "what is not" is restricted to (one or both of) the two senses of *to mê on haplôs* analysed at *GC* 317b5–13. It is not until Aristotle introduces a third way of analysing unqualified non-being in *GC* I 3 that conceptual space opens up for thinking about how it is possible for substances to come into being in the unqualified sense.

What, then, of *Physics* 191b27–9, where Aristotle appears to anticipate the *GC* solution as 'another way' of resolving the Eleatic Challenge? Does that not show that he was aware of that analysis in *Physics* I 8? And if so, how do we explain the difference between their accounts of unqualified becoming? The first thing to note is that the reference to an alternative solution at *Physics* 191b27–9, which consists in drawing a distinction between potentiality and actuality, may not be a reference to *GC* I 3 after all. For *GC* I 3 also appears to attribute that distinction to another work. Aristotle first mentions the distinction at 317b16–17, which he seems to include among those distinctions that have already been established elsewhere (b18). He then says that, even with these distinctions in hand, it is still 'extraordinarily difficult to see how there can be unqualified becoming – whether we assume that it proceeds from what is potentially or in some other way'. And so, he says, one might still raise the puzzle (*aporêseie*) about 'whether substance, in the sense of a "this something", comes to be at all' (b18–22). So it may be that the innovation of *GC* I 3 is not the application of the distinction between potentiality and actuality to the problem at hand but its insight into how to characterise the potential being from which substantial generation proceeds (317b23–33). In that case the reference to an alternative solution at *Physics* 191b27–9 may not be enough to connect *Physics* I 8 to the *GC* I 3 account.

Even if the reference is to *GC* I 3 the fact remains that *Physics* I 8 denies, while *GC* I 3 accepts, that things can come into being from what is not in the unqualified sense (191b13–14). Now I agree with Algra that *Physics* 191b13–14 may simply mean to rule out generation *ex nihilo*. And on that Aristotle's position never changed. But there is no hint in *Physics* I 8 that the distinction between potentiality and actuality is somehow meant to qualify or temper *that* claim. The denial in *Physics* I 8 that anything comes to be from what is not in the unqualified sense appears to be absolute. Nor is there any suggestion that the analysis at *Physics* 191b13–16, which associates non-being with the privation, is restricted to cases of *qualified* becoming, as it clearly is in *GC* I 3. There Aristotle says that 'coming to be something proceeds from what is not something, e.g. from the not-white or the not-beautiful' whereas 'coming-to-be in the unqualified sense proceeds from what is not in the unqualified sense' (317b3–5). The bottom line is that *GC* I 3 accepts both kinds of changes into its ontology whereas *Physics* I 8 only recognises the former. That is a striking difference.

In light of all this I would suggest that the reference to another solution at *Physics* 191b27–9 is either a reference to some other work aside from *GC* I 3 or else it is a later addition to the text either by Aristotle himself or by a later editor. Given the disagreements between *Physics* I 8 and *GC* I 3 just outlined, both of these strike me as reasonable bullets to bite.

As we shall see in Chapters 2 and 3, this is not the only place where Aristotle's views on generation appear to have developed between *Physics* I and *GC* I. It is generally held that in *Physics* I 7 Aristotle argues that all forms of coming-to-be, including substantial generation, meet what I shall call the Persistence Condition, which says that when something comes to be the subject survives the change to become a constituent of the finished product. Yet Aristotle categorically denies this in *GC*. There he argues that what distinguishes coming-to-be in the unqualified sense from all other forms of change is precisely the fact that the subject does not persist through the change. If this is right, then this gives us another reason to think that *GC* represents a further stage in Aristotle's thinking about the core cases of generation.

CHAPTER 2

The Hylomorphic Model of Substantial Generation: Physics *I*

At the outset of the *Physics* Aristotle tells us that when the objects of any inquiry have 'principles, causes, or elements', knowledge and understanding are acquired by grasping these: 'For we do not think we know a thing until we are familiar with its primary causes or first principles and have carried out our analysis as far as its elements. Clearly, then, also in the science of nature our first task will be to determine its principles' (184a10–16). The aim of the next three chapters is to present Aristotle's general model of substantial generation by examining those principles from which natural substances first come into being. Each of the three chapters focuses on a different set of texts. This chapter explores the basic hylomorphic model of coming-to-be from the perspective of *Physics* I, with an emphasis on the account in *Physics* I 7 (which is generally taken to be the *locus classicus* of that model). Chapter 3 turns to *GC* I 1–4, where Aristotle considers in more detail how substantial generation differs from changes in the non-substantial categories (e.g. alteration). Finally, Chapter 4 examines *GC* II 9, where Aristotle clarifies and extends the basic model from *Physics* I in several ways. In its most complete formulation, the hylomorphic model says that natural generation involves three principles: (1) matter, which is the subject from which the change proceeds; (2) form, which is the end towards which the process advances; and (3) an efficient cause, which (at least in the generation of living things) directs the process towards that form.

2.1 The Basic Model: *Physics* I 7

In *Physics* I Aristotle purports to discover the principles and causes of natural beings, from which they 'primarily are and come to be' (cf. 190b17–19). According to the basic model set out in *Physics* I every instance of coming-to-be (both qualified and unqualified becoming) involves something *S* that comes to be, and a pair of contraries *F* and *G*, which represent the two termini of the change. Within this tripartite structure, *S* is the subject (what Aristotle calls the *hupokeimenon*), while the contraries are identified with the form and its privation (e.g. 191a13–21; cf. *Metaph.* XII 2, 1069b33–5).[1] Together *Physics* I 5–7 argue that these three items (subject, form, and privation) constitute the basic principles of coming-to-be for all natural things: *Physics* I 5 argues that the contraries are principles; *Physics* I 6 shows that there must be some third thing besides these; and *Physics* I 7 argues that this third thing is the subject that underlies the change. Here I want to focus on the doctrine of *Physics* I 7.[2]

Physics I 7 begins by distinguishing various cases in which one thing comes to be from another in terms of whether their termini (the *terminus a quo* and *terminus ad quem*) are simple or complex. For example, we can say that a man comes to be musical (simple–simple) or the unmusical man becomes a musical man (complex–complex). According to Gill (1989, 99) Aristotle calls "simples" those items that are specified with reference to a single category (e.g. man [substance], unmusical [quality]) while "complexes" are composed of items from more than one category (e.g. unmusical man). After this preliminary discussion Aristotle goes on to make two claims about coming-to-be:

> Once these distinctions have been made one can grasp the following from all the cases of coming-to-be: that something must always serve as a subject, namely what comes to be, and that this, though always numerically one, is not one in form. By "in form" I mean the same as "in account"; for what it is to be a man is not the same as what it is to be musical. And one part survives, while the other does not: what survives is not an opposite (for the man survives), but not-musical or unmusical does not survive nor does the compound of the two, namely, the unmusical man. (*Ph.* I 7, 190a13–21)

The two claims in this passage are as follows. First, Aristotle tells us that all coming-to-be requires a subject (*to hupokeimenon*) from which the change proceeds, which is the very thing that comes to be (*to gignomenon*). Second, one part of the *terminus a quo* survives (the subject) while the other does not (the privation or the subject-cum-privation). I will call these two claims the Subject Condition and the Persistence Condition, respectively:

[1] In *Physics* I 5 Aristotle argues that everything that comes to be by nature (*ta phusei gignomena*) comes to be from its opposite or contrary (*enantia*) (188b25). He then goes on to argue in *Physics* I 7 that the contraries are the form and its privation (*hê sterêsis*). In *Categories* 10, however, Aristotle denies that possession/privation counts as a form of contrariety. I ignore this technical distinction here.

[2] For a general discussion of these chapters see Kelsey (2008, 2010) and Bostock (2006, ch. 1). Eventually Aristotle reduces the principles to two: form and subject. For a discussion of this see Ebrey (2018).

Subject Condition (SC): There must always be a subject from which a thing comes to be non-incidentally.

Persistence Condition (PC): That subject persists through the change to become a constituent of the finished compound.

Most scholars take matter to satisfy both conditions in the case of substantial generation. Indeed, many take *Physics* I 7 to be developing the concept of a subject in connection with the idea of being a continuant of change, so that SC and PC collapse into a single condition imposed on all coming-to-be alike. Bostock, for example, argues that Aristotle's expression *to hupokeimenon* is made to do 'double duty' in *Physics* I 7 for both the starting point of the change (the *terminus a quo*) and for what persists through the change (the continuant).[3] On this reading, to say that coming-to-be requires a *hupokeimenon* just means there must be something that persists through that change and survives as a constituent of the finished product. In contrast to this Charlton (1970, 135) points to *GC*, where Aristotle is careful to distinguish being a subject and being a continuant (see Chapter 3).[4] While Charlton argues that not all subjects of change persist through the change, at this stage in the argument I only want to make the weaker claim that in this context the concept of a *hupokeimenon* does not entail being a continuant. That point does not require taking a stand on the further question of whether or not Aristotle is committed to PC in all cases of change including substantial generation. Thus, in what follows I shall take *to hupokeimenon* to be playing only the single role of that from which the change proceeds (the *terminus a quo*). It is a separate claim to say that this subject also persists through the change to become a constituent of the finished product.[5]

It is obvious that cases of non-substantial change meet both conditions. For example, when Socrates comes to be healthy from being unhealthy there is a subject from which the change begins (Socrates), and that subject clearly persists through the change to become a constituent of the finished product (healthy-Socrates). Again, consider the process of learning where a child comes to be knowledgeable from being ignorant. In this case the

[3] Bostock (2006, 9). See also Broadie [Waterlow] (1982, e.g. 47), Gill (1989, 6, 90, 106), Witt (1989, 66).

[4] Charlton's argument is that not all subjects are continuants (as shown by *GC* I 1–4). But Aristotle's account of elemental transformation in *GC* I 4 also shows that not all continuants are subjects. For example, when water is transformed into air transparency survives, but qualities cannot be subjects (see Chapter 3).

[5] For a similar view see Kelsey (2008, 199–200, esp. n. 2), who also denies that persistence is part of the concept of being a *hupokeimenon* (though he follows the orthodox reading in taking PC to cover all changes).

child survives the transformation while her ignorance is destroyed and replaced by knowledge. Gill (1989) calls this the Replacement Model of coming-to-be, since it involves a persistent subject that replaces one property with another.

Things are less straightforward in cases of substantial generation. First of all, Aristotle is explicit that even this type of change must satisfy SC:

> "Coming-to-be" is said in multiple ways. Some things do not come to be [without qualification] but we say that a thing comes to be something (*ti gignesthai*). Substances alone come into being in the unqualified sense (*haplôs de gignesthai*). Accordingly, it is clear that in all cases other than substance there must be some subject that comes to be (*to gignomenon*) ... But that substances as well, and anything else there is that can be said to exist without qualification (*haplôs onta*), come to be from some subject will become evident upon further examination. For there is always something that serves as the subject from which what comes into being emerges, e.g. plants and animals come into being from seed (*ek spermatos*). (*Ph.* I 7, 190a31–b5)

In this passage Aristotle tells us that in both qualified and unqualified becoming there is always a pre-existing subject from which the change proceeds. In the case of living things, this role is played by the seed (e.g. an oak tree comes into being from an acorn).[6] What is not clear, however, is whether or not Aristotle means to commit himself to the further view that when a thing comes to be in the unqualified sense this subject also persists through the change to become a constituent of the resulting substance. This forms one of the central issues of the next two chapters. Before turning to that, let me first say something about the Subject Condition.

2.2 The Subject Condition

According to SC all coming-to-be requires an underlying subject from which the change proceeds. The subject is said to be numerically one but

[6] The concept of *sperma* plays an important role in Aristotle's theory of biological generation (see Chapters 5 and 6). There is no satisfactory English translation of the Greek that covers all its various uses. In Aristotle's own biological writings it is often used simply as a general term for the "reproductive residue" (*perittôma gennêtikon*) of an organism. Other times it is used as a more specific term to refer to male semen (technically *gonê*: e.g. 727b34), sometimes to female menstrual blood (technically *katamênia*: e.g. 728a26–7, b23, 750b4–5, 767b16–17, 771b20, b22–3), and sometimes for the embryo that results from their interaction (technically *kuêma*: e.g. 724b14–18, 728b34–5; cf. 731a2–4). For the most part I adopt the generic term "seed" (which includes plant seeds) and remain neutral on what Aristotle means to indicate by the term. When it is important to be more precise, I will use a more descriptive translation.

not one in form (cf. 190b23–5). For example, when a musical man comes to be from an unmusical man, both "man" and "unmusical" pick out one and the same thing (e.g. Socrates). But what it is to be a man differs in account from what it is to be unmusical (they have different definitions).[7] According to *Physics* I 7 what comes into being comes to be *per se* from the subject (man) but only *per accidens* from the privation (unmusical) insofar as the privation coincides with that subject (190b16–20, 190b24–7).[8]

Aristotle's first argument for SC comes in the context of those cases where a thing comes to be in the qualified sense (190a33–b1). He has said that we sometimes say things like 'what is not musical comes to be musical' or that 'from unmusical he becomes musical'. However, in these cases what is coming to be – what undergoes the change – is not the contrary itself since contraries cannot exist in their own right but are always predicated of some subject: 'In all cases other than substance it is clear that there must be some subject (*hupokeisthai ti*), which is the thing that comes to be (*to gignomenon*). For when a quantity or quality or relation, time, or place comes to be there is something that underlies them since substance alone is not predicated of any other subject' (cf. 189a27–32). Aristotle's point here is that the contraries themselves cannot be *ta gignomena* since they are incapable of separate existence; instead, there must be something else that underlies them as a subject. Note that this is not an argument for the claim that only substances come to be without qualification (190a32–3). As I argued in the Introduction, what Aristotle means to rule out there is the possibility that accidental compounds might come to be in the unqualified sense. For example, we do not say that the pale man comes to be without qualification but only that the man comes to be pale (cf. *GC* 318a34–5, 319a3–16). Instead the present argument is meant to show that the contraries themselves do not come to be in any sense – with or without qualification – but rather there must always be some substance that underlies them as a subject, which is what comes to be.

According to the first argument, then, when a thing comes to be in the qualified sense (e.g. when the unmusical comes to be musical) what

7 Gill (1989, 100–1). For a more robust reading of this claim see Kelsey (2010, 118 n. 15).

8 According to Kelsey (2010) this alone constitutes the thesis Aristotle defends in the first half of *Physics* I 7, namely, that all coming-to-be involves a subject from which things come to be non-incidentally and that this subject is numerically one but not one in form. This two-part thesis, he argues, 'is *the* moral Aristotle draws from the observations he collects about how we speak about coming to be (190a13–16)' (Kelsey 2010, 112). It is worth noting that, while Kelsey thinks this implies that something persists through the change, he does not think Aristotle actually argues for that claim anywhere in *Physics* I 7 or that it forms part of the central argument of the chapter.

undergoes the change is not just a bundle of properties, since properties cannot exist in their own right. Rather, there must be some other thing besides the contraries that underlies them as a subject (cf. *Ph.* I 6). Aristotle then turns to cases where a substance comes to be without qualification. The argument here is more empirical. When we consider the standard cases of substantial generation – the generation of living things – we always *see* a pre-existing subject from which the new substance comes to be, namely, a seed. In this way, both arguments are meant to establish that coming-to-be always proceeds from some pre-existing thing that meets the conditions for being an ultimate subject of predication.[9]

We can now turn to the Persistence Condition. The question is whether Aristotle thinks all forms of coming-to-be, including substantial generation, involve an enduring subject or whether this is true only in cases where *S* comes to be *F* in the qualified sense.[10]

2.3 The Persistence Condition

We can begin by carving up interpretations of *Physics* I 7 into two main camps. According to the standard reading, *Physics* I 7 argues that in all coming-to-be something must persist (as one and the same subject) through the change and that in cases of substantial generation this role is played by matter. This reading is by far the most widespread in the literature.[11] By contrast, some commentators take PC to apply only to changes in the non-substantial categories.[12] I will call this "the restrictive

[9] Kelsey (2008, 203). Compare *GC* I 3, 317b8–11.

[10] The following two chapters significantly expand on Henry (2015a). The arguments here supersede those of that earlier paper, though the main thesis remains substantially the same.

[11] The standard interpretation is defended in some form or another by Broadie [Waterlow] (1982, e.g. 46–7), Gill (1989, 6, 98–108: discussed below), Code (1995, 415–17), Bostock (2006, 8–9), and Kelsey (2008, 2010, e.g. 112). Some remarks on the literature are in order here. Bostock (2006, 8) is careful to restrict his account to *Physics* I and does not speculate as to whether Aristotle endorsed PC outside the *Physics* (he clearly has *GC* in mind: see below). Broadie [Waterlow] (1982) endorses the standard reading and takes the concept of underlying subject to entail the idea of persistence (e.g. 47). But she seems to soften this position in Broadie (2004) (again in light of the evidence from *GC*) by claiming that *Physics* I 7 is actually silent on the issue (129–30). Kelsey (2010) parses *Physics* 190a13–21 slightly differently so that PC is not at issue (see note 8 above). Finally, Charles (2018) takes *Physics* I 7 to commit Aristotle only to the claim that in substantial generation the thing from which generation proceeds underlies the entire process, but it need not remain the same sort of thing at all stages in the process. For example, what remains through the generation of a plant was initially a seed, but it need not remain a seed at every subsequent stage in the process (though see Charles 2004).

[12] Charlton (1970, 75–7, 133–5) is the most explicit defender of this position. He is motivated in part by a desire to reject the concept of prime matter and in part by the fact that in *GC* Aristotle is explicit that 'if anything did remain in all cases, there would be no such thing as coming into existence, but

reading". While some defenders of the restrictive reading hold that *Physics* I 7 denies that substantial generation involves a persistent subject (e.g. Charlton 1970), in this chapter I only want to defend the weaker thesis that the evidence for the standard reading of *Physics* I 7 is circumstantial and that it is possible to read that evidence in a way that is at least compatible with a version of the restrictive reading. According to this weaker version, *Physics* I 7 is best viewed as undeveloped with respect to the Persistence Condition. While Aristotle is explicit that the subject survives in those cases where a thing comes to be in the qualified sense ('the man survives, but not-musical or unmusical does not survive, nor does the compound of the two, namely, the unmusical man'), he does not reflect on, let alone argue for, the claim that this also holds for substantial generation. For example, when Aristotle turns to substantial generation at 190a31–b5, while he is explicit that it requires a pre-existing *hupokeimenon*, he does not say that the *hupokeimenon* persists through the change.[13]

Admittedly, if we restrict ourselves to the text of *Physics* I itself we can easily read it as supporting the standard reading. However, there are two reasons I raise this alternative reading here. First, if the interpretation of *GC* defended in Chapter 3 is correct (and it is certainly the most natural), then the standard reading of *Physics* I 7 comes at a price. Aristotle is explicit in *GC* that when a thing comes to be in the unqualified sense the subject does *not* survive. This forms the main axis of division for distinguishing substantial generation from changes in the non-substantial categories. This does not sit well with the standard reading of *Physics* I. For it implies that Aristotle changed his mind on the Persistence Condition between *Physics* and *GC*. Bostock seems to have this very worry in his sights when he says:

> One might also sympathize with Charlton's claim . . . that there are several passages elsewhere, especially in the first book of the *De Generatione et Corruptione*, which are (as he puts it) "not propitious" for interpreting Aristotle as claiming there that there is something which persists when (say) air changes into water or water into earth.[14] But the question at issue is whether Aristotle does, in *Physics i*, claim that there is always something that persists through any change, and it seems to me that the text is quite unambiguous on this point. (Bostock 2006, 8)

only alteration' (77). See also Jones (1974) and, with qualification, Broadie (2004). Frey (2007, cf. 2015) also denies that matter persists in substantial generation, which forms the basis of his response to the so-called Ackrill Problem.

[13] See Broadie (2004, 129–30).

[14] Charlton (1970, 77) is actually more forceful on this, noting that according to *GC* 'if anything did remain in all cases, there would be no such thing as coming into existence, but only alteration'.

One obvious virtue of adopting the more restrictive reading of *Physics* I 7, then, is that it avoids this worry since on that reading Aristotle only means to commit himself there to the weaker claim that substances come to be *from* some pre-existing subject and not the stronger claim that this subject also *persists* through the change.

Second, Aristotle shows absolutely no concern with the issue of persistence in the *GA*. This is revealing. For if Aristotle did have strong reasons for positing a continuant of substantial generation, we should have expected that thesis to form a significant part of his account there. For he treats living things as the paradigm natural substances and biological generation as substantial generation par excellence. Yet, not one of the puzzles raised in the *GA* is devoted to the question of what exactly is the persisting subject – despite all the dust that has been kicked up by modern commentators who have tried to problematise the issue (e.g. Freeland 1987, 392).[15]

These two considerations provide us with a reason for adopting a more restrictive reading of PC at least outside of the *Physics*. In light of this it will be useful to have on the table an alternative interpretation of *Physics* I 7 that is compatible with that external evidence. The aim of the next two sections is not so much to argue against the standard reading of *Physics* I 7 as it is to motivate this alternative, which is more in line with Aristotle's views elsewhere. I do this by suggesting that the text of *Physics* I is not as unambiguous on this point as Bostock claims. In the end, however, I leave it to the reader to assess the evidence of *Physics* I and decide which way of reading the text makes the best sense of Aristotle's views there.

2.4 Re-Evaluating the Standard Reading

I want to begin by first addressing Bostock's claim that in *Physics* I 7 the concept of a *hupokeimenon* is made to do 'double duty' as that from which the change proceeds and what persists through the change. Aristotle never actually links the two ideas in this way. For example, neither of his arguments for SC invokes the idea of persistence. While it is true that Aristotle thinks in every change what a thing comes to be from (the

[15] See Chapter 3, Section 3.5.1. Connell (2001, 307–12) also takes note of this fact about the *GA*. However, she appears to accept the orthodox reading of *Physics* I 7 and takes the lack of concern with persistence in the *GA* as evidence that Aristotle's account of substantial generation in the biological realm does not conform to the abstract model from the *Physics*. By contrast, I see this lack of concern in *GA* as a reason to question the orthodox reading of *Physics* I 7.

terminus a quo) is also an ultimate subject of predication, he makes a similar claim in *GC* (317b8–11). And yet, he is categorical there that substantial generation does *not* involve a persistent *hupokeimenon*. So clearly Aristotle does not think persistence simply falls out of the idea that all change requires a subject that also satisfies the conditions for being a subject of predication.

Defenders of the standard reading typically cite two other pieces of evidence from *Physics* I 7 to show that Aristotle thinks all changes, including substantial generation, involve a persisting subject.[16] First, there is the passage at *Physics* 190a13–21 where persistence is first mentioned. Aristotle introduces that passage by saying that 'one can grasp the following from *all* cases of coming-to-be' (*hapantôn tôn gignomenôn*). Most take this as definitive evidence that coming-to-be in the unqualified sense is subject to PC.[17] And with good reason. Aristotle opens *Physics* I 7 with a methodological remark saying that the natural (*kata phusin*) method is to begin by surveying all coming-to-be, first marking off its common features (*ta koina*), and then going on to investigate what is peculiar to each form (*ta idia*). On one reading, 189b32–190a31 deals with what is common to all coming-to-be while 190a31 moves on to consider what is peculiar to different forms of the change.[18] Second, Bostock argues that the argument in *Physics* 190b10–23 entails that the subject remains even in cases of substantial generation:

> The only ground Aristotle could have for saying that whatever comes into being is composite (*sunthetos*) is that we can distinguish in it two "elements", one the persisting element (what underlies) and the other the acquired element (the form). If the element said to underlie did not persist in the end product there would be no ground whatever for saying that the end product was composite, and Aristotle explicitly claims that all products of becoming are composite.[19]

Let me start with the first piece of evidence. It is possible to read the opening sentence of 190a13–21 as restricting its scope to those cases of coming-to-be that Aristotle has just been discussing, namely, those where

[16] In addition to these, scholars also argue that Aristotle says in cases where a statue comes to be from bronze that the bronze survives (190a25–6) and that he uses this example to make a point about substantial generation. I return to this below. The other typical move that scholars make is to claim that PC allows Aristotle to avoid the Eleatic Challenge. But Aristotle never invokes the idea that something persists in his response to Parmenides in *Ph.* I 8 (Kelsey 2006, 200–1; Anagnostopoulos 2013, e.g. 273).

[17] Bostock (2006, 8), cf. Broadie (2004, 129–30), Gill (1989, 98, 100).

[18] Gill (1989, 100–1).

[19] See also Kelsey (2008, e.g. 200 n. 22).

a thing comes to be in the qualified sense (189b32–190a12).[20] (This is at least the sense of the Oxford translation.) And all of the examples Aristotle has considered thus far are of non-substantial changes. More importantly *Physics* I 7 does not present any argument for the (surely controversial and not obvious) claim that even in substantial generation something persists through the change.[21] When Aristotle summarises what is to be gleaned from cases of substantial generation at 190b9–10 he says only that 'it is clear for all things that come to be in this way that they come to be *from* a subject' (*ex hupokeimenon gignetai*). He says nothing about whether or not that subject also *persists* through the change.[22]

Let me now turn to Bostock's claim that *Physics* 190b10–23 entails that the subject remains in cases of substantial generation. This is certainly an interpretative move on his part. Bostock reads this as an *argument* for hylomorphism and so takes PC as grounds *for the conclusion* that natural substances are composites of matter and form. But it isn't obvious that 190b10–23 is an argument. Aristotle says that it is clear 'from what has been said' that what comes into being is always complex. And what has just been said is that all cases of coming to be involve a subject from which the change proceeds (190b9–10). Since Bostock assumes that the *Physics* I concept of a *hupokeimenon* already includes the idea of persistence (an assumption we have already called into question), he takes 190b10 as a straightforward claim that what comes into being is complex *because* the subject persists through the change. But that is not the only way to read the passage. Aristotle's claim at 190b10 is likely just a reiteration of the earlier conclusion drawn from the analysis of the ways *we speak* about coming-to-be, namely, that the idea that what comes to be is complex is already implicit in language (189b32–190a4). The bulk of the discussion thus far can thus be read as an attempt to unpack that implication with more precision using his own distinction between subject and contrary. On this reading all 190b10 need be saying is that, when we reflect on how people speak about change, it becomes evident that what comes to be is complex. Indeed, when we say the non-musical man becomes a musical man it is evident that both what comes to be and what it becomes are

20 Cf. Bostock (2006, 8).

21 Kelsey (2010, 114 n. 9) suggests that Aristotle would not have found it controversial to say that in every change the subject persists. However, in *GC* I 1 Aristotle tells us that this very issue is what divided those who reduced generation to alteration from those who treated them as distinct phenomena. So apparently it was controversial.

22 Charlton (1970, 77). Gill (1989, 103) concedes that Aristotle does not say here that any part of the pre-existing subject persists in substantial generation. But she thinks 'this omission need not indicate that nothing survives'. True, but it doesn't imply that either.

complex (190a3–5). Although Aristotle supports this conclusion by drawing on cases where the subject survives the change (190a22–6), his general point in no way depends on that.[23]

So the evidence for PC in *Physics* I 7 turns out to be a lot more circumstantial than defenders of the standard reading make it out to be. While Aristotle does argue that even when a thing comes to be in the unqualified sense there must be a subject from which the change proceeds, he does not make the stronger claim that this subject also survives to become a constituent of the finished product. Instead the standard reading relies on inference to connect Aristotle's various remarks to the conclusion that substantial generation involves a persistent subject.

Finally, it is worth noting that even defenders of the standard interpretation concede that *Physics* I 7 is not as explicit about the Persistence Condition as it has traditionally been made out to be. For example, Kelsey (2010) argues the central thesis Aristotle defends in the first half of *Physics* I 7 is (a) that all coming-to-be involves a subject from which things come to be non-incidentally and (b) that this subject is numerically one but not one in form. While he does think *Physics* I 7 comes 'close enough' to saying that matter survives in substantial generation, he concedes that Aristotle never explicitly defends that thesis there.[24] Indeed, he agrees that that claim forms no part of the central thesis of the chapter or in any way contributes to it. Gill (1989) is similarly cautious about her conclusions. She holds that the so-called Replacement Model developed in *Physics* I 7 is not presented there 'as the scheme for all changes' but is instead a model for coming-to-be only in the qualified sense (Gill 1989, 98, 108). Nevertheless, she thinks that certain aspects of that model are meant to be generalisable to substantial generation, including the fact that change involves three fundamental principles and that one of these (the subject) persists through the change. However, she also concedes that Aristotle omits to say in that chapter that any part of the subject survives in cases of substantial generation (Gill 1989, 103).

2.5 *Physics* I 9

Although my account has focused mainly on *Physics* I 7, there are two passages in *Physics* I 9 that remain problematic for the restrictive reading:

[23] Cf. Kelsey (2010). By contrast Michael Ferejohn (personal correspondence) argues that persistence is in fact at issue in the linguistic analysis, which he takes to apply to substantial generation as well. See also Charles (2018).

[24] Kelsey (2010, e.g. 112–14).

(1) 'Their [sc. the Platonists'] triad is therefore of quite a different kind from ours. For they got so far as to see that there must be some underlying nature, but they make it one – for even if one philosopher [sc. Plato] makes a dyad of it, which he calls Great and Small, the effect is the same; for he overlooked the other nature. For the one which persists (*hê hupomenousa*) is a joint cause, with the form, of what comes to be – a mother as it were.' (192a9–14, Hardie and Gaye translation)

(2) 'For if it came to be, then something must have existed as a primary substratum from which it should come to be and which should persist in it; but this is its own very nature, so that it will be before coming to be. For my definition of matter is just this – the primary substratum of each thing from which it comes to be and which persists in the result not accidentally.' (192a29–32, Hardie and Gaye translation)

Defenders of the standard reading will take 'what persists' in (1) as shorthand for Aristotle's concept of matter, which is confirmed by (2) where he apparently defines matter in part by its ability to persist through the change.[25]

I do not think (1) is decisive, since Aristotle makes that statement in the context of discussing Platonism. So, it is not clear whether 'the one which persists' is shorthand for his own technical concept of matter or for some item in the theory under consideration (e.g. the receptacle of the *Timaeus*). Of course (2) as translated does commit Aristotle to the standard reading. But this is not what Aristotle actually says. The Greek word that gets translated as 'persists' in (2) is the participle form of *enhuparchein* (not *hupomenein*), which in this context means to be inherent in a thing.[26] Persistence is stronger; it implies being inherent in the thing all the way through the change and not just at the start. With this in mind we can retranslate 192a31–2 as:

(3) 'For if it came to be, then something must have existed as its primary subject from which it comes to be as something inherent in it; but this is its own very nature, so that it will be before coming to be. For my definition of matter is just this – the primary subject of each thing from which it comes to be, which is inherent in it non-accidentally.'[27]

[25] Gill (1989, 148).

[26] This is also the word that Hardie and Gaye translate as "persists" in the account of the material cause at *Physics* I 3, 194b24 (which betrays a preference for the orthodox reading).

[27] This is similar to how Charlton (1970) translates the passage.

This does not obviously commit Aristotle to PC in cases of substantial generation. If I am right, then whether or not the material principle also survives the change is not relevant to Aristotle's point here. Instead, I propose we read the passage as follows.

First, when Aristotle says that matter is the primary subject from which a thing comes to be 'which is *inherent in it* non-accidentally', at least on one interpretation he means inherent in the subject that forms the starting point of that change (e.g. the seed) not the substance that results at the end of the change (e.g. the adult organism). For example, after arguing that females provide the matter for generation (*GA* I 19–20), Aristotle opens *GA* I 21 by asking how semen from the male is a cause of the thing that comes into being (*tou gignomenou*): 'Does it serve as a cause by being straightaway an inherent part (*hôs enhuparchon kai morion*) of the body of the thing coming to be, mixing together with the matter from the female? Or does the bodily portion of the semen have no share in generation but only the power and motion in it?' (729b1–6) The question here is about the causal role that semen plays in the formation of the embryo, which is the subject that comes to be the animal at the end of the change.[28] When he asks whether semen acts as a cause of generation by being 'an inherent part of the body of the thing coming to be' he is asking whether it forms a material constituent of the embryo, not whether it *persists* through the change to become a constituent of the animal that results from it. Persistence is not at issue here.

In light of this, I would suggest that by referring to matter as something 'inherent' in the thing that comes to be Aristotle is not attempting to characterise its role as the *persisting element* of the change. Instead, his point is to further clarify the sense in which matter functions as that 'from which' (*ex hou*) a thing first comes into being. For we have just seen that there are two principles that count as the *ex hou* of generation – the material principle (e.g. menstrual blood) and the moving principle (e.g. semen) – only the first of which makes its contribution to generation by being 'inherent in' the thing that comes to be. The other (the agent or moving cause) does not provide anything material to the change but instead effects generation solely by virtue of the active power and motion it contains. If this is right, then (3) does not commit Aristotle to the claim that the very essence of matter is to be a continuant of change. That is not the point of (3).

[28] See Chapters 5 and 6. Aristotle often uses *to gignomenon* for the subject (*to hupokeimenon*) that undergoes the change (the *terminus a quo*) (e.g. *Ph.* 190a15).

The aim in this chapter was to lay out the basic hylomorphic model of natural generation as presented in *Physics* I 7, which specifies the role of matter and form (and privation) in the generation of a new substance. According to this model, matter is the subject from which the change proceeds (the *terminus a quo*) while form is the positive state that is acquired as a result of that change which tells us *what* is coming to be, e.g. a human or a horse (190b10–23).[29] But this is not the full story. I shall argue that *Physics* I presents only the bare bones of the hylomorphic model of generation (what Myles Burnyeat [2004, 12] calls 'the less sophisticated triadic model of matter, form, privation'). Aristotle will go on to argue in *GC* II 9 that matter and form alone are not sufficient to explain natural generation but that we must have recourse to a third principle over and above these, namely, a primary efficient cause. Moreover, the bare bones model tells us nothing about the role of these in a dynamic teleological context such as the development of living things. I shall explore these two extensions of the hylomorphic model in subsequent chapters (especially Chapters 4 and 8). But first I want to turn to the account of substantial generation in *GC* I 1–4, which provides more support for what I have called "the restrictive reading" of the Persistence Condition.

[29] Compare *Metaph.* 1032a12–25, 1033a24–8, and 1070a1–2 (*ho de hê hulê, eis ho de to eidos*).

CHAPTER 3

Substantial versus Non-Substantial Change: GC I 1–4

In Chapter 2 we raised the question of whether Aristotle thinks that in all cases of coming-to-be (including substantial generation) the subject from which the change proceeds must persist through that change or whether he thinks this holds only for those cases where *S* comes to be *F* in the qualified sense. I tried to show that, if Aristotle did think that in cases of substantial generation the pre-existing matter survives to become a constituent of the finished product, it is not obvious that he provides that defence in *Physics* I as the standard reading assumes. However, because the idea of persistent matter has such a firm hold on the literature, much more will be needed to loosen its grip. Part of the aim of the current chapter is to provide that additional support by examining the account of generation in *GC* I 1–4. According to this account, the subject from which generation proceeds does not survive (even potentially) in the finished product but is completely transformed in the course of becoming a substance. As Aristotle puts it: the subject is transformed 'as a whole', which he glosses as a change in its matter and form together. This is ultimately what distinguishes generation from changes in the non-substantial categories.

3.1 Generation versus Alteration

Generation and Corruption opens by setting out the aims of the inquiry. Among the central questions to be examined is whether generation and alteration 'have the same nature' or whether 'they are different names corresponding to separate processes' (314a6–8; compare the opening of *GC* I 3).[1] What Aristotle is asking here is whether all putative cases of generation are reducible to alteration or whether generation can be

[1] Aristotle typically uses "alteration" (*alloiôsis*) for change in quality, which he distinguishes from growth (change in quantity) and locomotion (change of place), as he does (e.g.) at *GC* 314a4–5. But throughout *GC* I he tends to use the term more broadly (as he does here) to stand in for all types of non-substantial change (Brunschwig 2004, 33–4). Thus, unless otherwise stated, I shall use

considered a distinct change in its own right. When a seed develops into a mature organism, is this a genuine case of something new coming into being that was not there before or is it simply the modification of some pre-existing element that remains one and the same throughout the change? Aristotle's answer to this question is that there is such a thing as coming to be in the strict and unqualified sense and that this sort of change is *sui generis*. It is not reducible to a change in one of the other non-substantial categories (*GC* 319b6–8). Establishing this fact is the main project of *GC* I 1–4.

Aristotle begins the inquiry by dividing earlier views on this question into two camps. The monists are said to have reduced generation to alteration, while the pluralists maintained a distinction between them:

> It is necessary for those who construct all things out of a single element to say that coming-to-be and passing-away are alteration. For [on their account of generation] the underlying subject always remains one and the same, and we call that sort of change "being altered". On the other hand, those who make the ultimate kinds more than one make alteration different from generation. For [on their account] generation and destruction occur as things come together and break apart. (*GC* I 1, 314b1–6)

Notice the role played by the Persistence Condition in the way Aristotle frames this debate. The monists are said to have reduced generation to alteration precisely because they held that in all coming-to-be there is a single element that serves as a persistent substratum (cf. 332a8–9; *Metaph.* I 3, 983b6–18). As we shall see, *GC* I 1–4 treats the presence or absence of a persisting subject as the main axis of division between alteration and generation. But first let me say something about the other party to this debate.

At first glance it isn't obvious that the pluralists did distinguish generation from alteration as Aristotle suggests. By his own lights, some of the pluralists reduced it to alteration while others were eliminitivists about generation. At 314a13, for example, Aristotle says that Anaxagoras 'failed to understand his own utterance; for he *says*, at any rate, that coming-to-be and perishing are the same as alteration'. Empedocles is said to have denied that there is such a thing as coming-to-be 'but only a mingling and divorce of what has been mingled' (314b6–8). Finally, Democritus not only denies that the ultimate constituents of reality, the atoms, are generated and destroyed; he denies that when they come together to form compound

"alteration" to represent all forms of coming-to-be in the qualified sense, while "generation" will continue to stand for coming-to-be in the strict and unqualified sense.

bodies anything new comes to be as a result (Aristotle *ap.* Simplicius, *in DC* 295,9). Thus, in Democritus' world, nothing comes to be or passes away in the strict and unqualified sense.[2] Given the dichotomy of positions in *GC* I 1, surely it would be more natural to think of these pluralists as having treated putative cases of generation as the alteration of underlying elements that remain numerically the same throughout the process. So why does Aristotle place them in that other camp?

Note, first, that what Aristotle actually says is that the distinction between generation and alteration is an *implication* of their view ('they *must* maintain', 314b5–6; 'they *must* recognise', 314b11) not something they necessarily defended. Indeed, Aristotle admits that it is difficult for some pluralists (notably Empedocles) to maintain the distinction between generation and alteration consistently. Nevertheless, we can see why he classifies them as he does if we keep in mind the framework for distinguishing generation and alteration at work in *GC* I. Atomism provides a nice illustration of this. Imagine a group of atoms are randomly joined together to form what common sense refers to as a tree (a new compound). Over the course of its lifespan those atoms remain conjoined and change only in their internal configurations relative to one another, until eventually they get separated out again and the compound is dissolved. If we focus on the atoms themselves, then nothing comes to be or passes away; atoms (which are eternal) change only in respect to their position and orientation. From this perspective it might be more natural to think of this as an alteration of the underlying atoms that persist through the process. However, if we think of *the whole underlying cluster of* atoms as a subject of change in the *Physics* I 7 sense, then given the framework of *GC* I we can see why Aristotle finds it reasonable to classify this as a case of generation in contrast to alteration. From this perspective we can distinguish the combination and dissolution of the atoms, on the one hand, from the changes that befall that enduring cluster, on the other. When the atoms are combined to form a new cluster we say that a tree has come to be, and when those same atoms are separated out again we call that a destruction and say that the tree has been destroyed. By contrast when that specific cluster of atoms (the underlying subject) remains one and the same and changes only in its internal

[2] Broadie (2004, 127) claims that Aristotle means to treat atomism as a theory that 'reduces all macroscopic change to non-substantial change (locomotion and turning) in imperceptible subjects' in this way. That may have been Democritus' intention, but Aristotle clearly treats atomism here as a version of the pluralist model that distinguishes generation from alteration. So Aristotle at least thinks atomism does not entail a reduction of generation to alteration. Nevertheless, I am sympathetic to Broadie's assumption precisely because it is natural to think of atomism in that way.

configurations, we can call that sort of change an alteration of the tree. Thus, even if Democritus himself would not make the distinction between these two sorts of change, it is not hard to see why Aristotle thinks his view at least implies such a distinction.

In *GC* I 2 Aristotle tells us that the pluralists' way of distinguishing generation and alteration is bound up with their particular model of coming-to-be. He introduces this model at *GC* I 2, 315b15–19:

> Since nearly all our predecessors think that generation and alteration are different and that generation and destruction involve aggregation and separation (*sungkrinomena kai diakrinomena*) while alteration involves changes among a thing's affections (*metaballontôn tôn pathêmatôn*), we must concentrate our attention on these theses.

According to this model generation and destruction of macroscopic bodies are effected through the aggregation and separation of the basic elements, while alteration involves changes in their incidental properties. While Aristotle spends most of *GC* I 2 picking apart the arguments for atomism, his ultimate goal in that chapter is to undermine this alternative model in order to clear the way for his own preferred account of generation (see T1 to T3 below). At first glance it isn't immediately obvious how the question about indivisible bodies in *GC* I 2 is related to the model introduced at 315b15–19. According to Sedley (2004, 89) Aristotle picks on atomism because he thinks it represents the strongest version of that model and thus seeks to show that even in its best form that model fails. But this overlooks Aristotle's own remarks about the connection between atomism and the alternative model under consideration in *GC* I 2. At 315b24 we are told that the evaluation of that model ultimately turns on whether or not there are indivisible magnitudes. Aristotle then says that '*if* coming-to-be and passing-away really are constituted by aggregation and separation, then a body *must* contain atomic magnitudes that are indivisible' (316b32–4).[3] So Aristotle zeros in on atomism, not because he thinks it is the strongest version of the model under consideration, but because he thinks that model *requires* the existence of atoms. Thus if Aristotle can show that atomism is false (there are no indivisible bodies: 317a1–16), then we have good reason to reject the pluralists' way of distinguishing generation from alteration, which then clears the decks for Aristotle's own preferred model.

[3] Aristotle argues that, if there are indivisible magnitudes, then they must be indivisible bodies (i.e. atoms) since the Platonic idea of indivisible planes is untenable (315b30–316a14).

It is not entirely clear why Aristotle thinks that the model under consideration in *GC* I 2 entails atomism. What exactly is the connection between saying that generation and destruction are effected by the amalgamation and separation of underlying elements, on the one hand, and the claim that these elements must be indivisible magnitudes, on the other? One possibility is that Aristotle thinks atomism provides the only plausible foundation for the alternative model. That argument might run something like this: (i) generation involves the amalgamation of ultimate constituents; (ii) denying the atomic thesis entails that these ultimate constituents are non-magnitudes, either size-less points or absolutely nothing; (iii) but you cannot build up a magnitude by putting together non-magnitudes of either sort; therefore (iv) if generation and destruction are effected by the amalgamation and separation of ultimate constituents, then those constituents have to be atoms. Even if this is not his reasoning, the structure of the argument in *GC* I 2 is clear. Since Aristotle thinks the alternative model (somehow) entails atomism, then showing that atomism is false gives us grounds for rejecting that alternative model.[4]

3.2 The *GC* Model of Substantial Generation

While Aristotle agrees with the pluralists that 'alteration involves changes among a thing's affections', he rejects the idea that generation and destruction are due to the aggregation and separation of some more basic underlying elements. He introduces his alternative model of generation in a key passage at the end of *GC* I 2:

> T1 Coming-to-be and passing-away in the unqualified and perfect sense (*hê haplê kai teleia genesis*) are not, as some assert, defined by aggregation and separation while alteration is a change in what is continuous. On the contrary, this is where the whole mistake lies. For coming-to-be and passing-away in the unqualified sense are not effected through aggregation and separation; rather, it occurs when *A* wholly changes into *B* (*metaballê ek toude eis tode holon*). But they suppose that all such change is alteration, whereas in fact there is a difference. For one component of the subject corresponds to the form and the other the matter: generation and destruction occur when there is a transformation in these; but when it is in its incidental affections it will be alteration. (*GC* I 2, 317a17–26)

[4] Philoponus (*in GC*, 24,20–25,7) is mistaken when he takes Aristotle to hold that atomism entails the alternative model but not vice versa, so that even if atomism is false the alternative model is still on the table. *GC* 316b3–4 is clear that *if* generation is constituted by the amalgamation of elements, then *there must be* atomic magnitudes. So, the structure of the argument is: *P* implies *Q*; ~*Q*; therefore ~*P*.

When we put this together with two other core passages the *GC* model of substantial generation begins to emerge:

T2 Since we distinguish between the subject (*to hupokeimenon*) and the affection whose nature it is to be predicated of that subject (*to pathos hô kata tou hupokeimenou legesthai pephuken*), and change occurs in each of these: It is alteration when the perceptible subject remains but changes in its own properties (the properties in question being either opposites or intermediates); for example, the body, although enduring as the same body, is now healthy and now ill, and the bronze is now spherical and now angular, and yet remains the same bronze. But when the thing is wholly transformed (*holon metaballê*) and nothing perceptible remains as its subject (e.g. when seed as a whole is converted into blood[5] or water into air or air as a whole into water) such an event is the coming-to-be of the one sort of thing and the passing away of the other, especially when the change proceeds from an imperceptible something to something perceptible (either to touch or to all the senses), as when water comes to be out of or passes away into air (for air is pretty imperceptible). (*GC* I 4, 319b8–21)

T3 Our account of growth must preserve the features of the subjects that are growing and diminishing. And these are three: first, any and every part of the growing magnitude is made bigger (e.g. if flesh grows, every particle of flesh gets bigger); second, it gets bigger by the accession of something; and third, it does so in such a way that the growing thing is preserved (*sôzomenou*) and persists (*hupomenontos*). For, whereas a thing does not persist (*oux hupomenei*) in the process of coming-to-be and passing-away in the unqualified sense, in both growth and alteration the growing subject itself and the thing that undergoes alteration does persist, though in alteration the quality, and in growth the size, does not remain the same. (*GC* I 5, 321a17–26)

The general model outlined in these three passages is the following (once again I will focus on generation). First Aristotle makes a distinction between the subject that undergoes the change, which is here said to be a complex of matter and form, and its incidental properties.[6] When the subject persists and changes merely in its incidental properties it counts as

[5] This example is puzzling since in Aristotle's own theory *gonê* (what I have translated as "seed") is typically reserved for the male semen. And he denies that semen is the subject that undergoes the change in animal generation. One option is to transpose "seed" and "blood", which would be consistent with Aristotle's theory in *GA* I 18 that says how blood is transformed into semen by concocting the leftover nutriment in the male's heart after digestion. Alternatively, we can take Aristotle to be talking about *gonê* in the sense used by some of his predecessors who do treat semen as the subject of generation. In that case the idea is that in generation semen is first transformed into blood, which is then transformed into tissues and organs.

[6] I take it that *to pathos* in T2 is meant to pick out 'the incidental affections' of the subject mentioned at the end of T1.

an alteration of that enduring subject. (This matches the way Aristotle characterises alteration in setting up the debate at 314b1–6.) For example, 'the body, although enduring as the same body, is now healthy and now ill, and the bronze is now spherical and now angular, and yet remains the same bronze'. Generation occurs when the subject does *not* persist but is completely transformed in the process, which Aristotle glosses (in T1) as being transformed in *both* its matter *and* form together. The contrast between the subject's matter and form, on the one hand, and its incidental affections, on the other, suggests that Aristotle is thinking of generation as a change in the deep ontological structure of the subject. This is why he characterises substantial generation as a *total* transformation of the subject.[7] Thus, whereas the monists maintained that the underlying subject always remains one and the same throughout change, on Aristotle's account the subject of generation does not remain, either formally or materially. It is only in cases where a thing undergoes a non-substantial change that the subject is said to be preserved (*sôzein*) and persist (*hupomenein*) as the very same subject. Indeed, T3 identifies the persistence of the subject as one of the very conditions for calling something growth (or alteration) *as opposed to* generation.

At the end of *GC* I 4 Aristotle says that a change counts as generation when nothing at all persists 'of which the resultant is a property or accident in any sense of the term' (319b34–320a2). This is just another way of saying that the subject does not persist (where *S* counts as a subject if it has other things predicated of it). The context of this remark is a worry about the ontological status of elemental transformation. When one element is transformed into another, at least one quality persists through the change. For example, when air (hot-moist) comes to be from fire (hot-dry) the hot survives, while dry is exchanged for moist. Given the way the *GC* model distinguishes alteration from generation one might worry that this counts as alteration rather than substantial generation; for in this case something does persists through the change (314b1–6). Aristotle's response is to point out that a quality cannot function as a proper subject. And so elemental transformation still counts as generation because, although something

[7] T1 allows that the subject that plays the role of pre-existing matter for one change is itself a composite of matter and form (cf. *Metaph.* VII 8). For example, in *GA* I 19–22 Aristotle argues that an embryo comes to be when semen from the male imposes shape and form on matter from the female. During development, this new subject undergoes a further change in its hylomorphic structure (it is transformed in matter and form together) resulting in a new substance coming to be. In this way the same thing (the embryo) counts as the final product of one change (embryogenesis) and the pre-existing matter of another (morphogenesis). I return to this at the end of Chapter 5.

persists, what persists is not an enduring *subject*. In our example, the moist (the resultant) is not *a property of* the hot (the continuant), since properties cannot underlie other properties as subjects. And so, *ex hypothesi*, the change does not qualify as alteration but generation (since the subject does not survive).[8]

3.3 Are *Physics* I and *GC* I Compatible?

Recall that in *Physics* I 7 Aristotle tells us that when one thing comes to be from another, the subject from which the change proceeds survives to become a constituent of the finished product (190a13–21). I called this the Persistence Condition (PC). In Chapter 2 I presented two ways of reading *Physics* I 7 with respect to PC. It should be clear by now that whether Aristotle's views on this remain consistent between *Physics* and *GC* depends on which of those two readings we adopt.

According to the orthodox reading, in *Physics* I 7 Aristotle means to say that all cases of coming-to-be, including substantial generation, require a persistent subject that endures (as one and the same subject) in the generated product. The problem with this reading is that it conflicts with Aristotle's stated position in *Generation and Corruption* I 1–4. According to that account, the presence or absence of a persisting subject is the main axis of division between generation and coming-to-be in the non-substantial categories. If the standard reading of *Physics* I 7 is correct, then it implies that Aristotle changed his mind about substantial generation. By contrast the restrictive reading holds that when Aristotle says in *Physics* I 7 that the subject of coming-to-be survives he is only thinking about those cases where a thing comes to be in the qualified sense (alteration, growth, locomotion). This reading does not entail that Aristotle changed his mind in *GC* since it takes *Physics* I 7 to be at best undeveloped with respect to that condition. While *Physics* I 7 holds that even substantial generation must proceed from a pre-existing subject (190b1–5) – a claim that forms the basis for his response to the Eleatic challenge in *Physics* I 8 (191a30–1) – Aristotle is not focused there on the further issue of whether this subject also persists through the change.[9] If this is right, then *Physics* I 7 is not incompatible with *GC*, as it would be on the standard reading, since Aristotle did not mean to commit himself in the *Physics* to any particular

[8] This is similar to how Broadie (2004, 130–3) reads the *GC* I 4 passage: while something persists in elemental transformation, it is not a persisting *subject*.

[9] Broadie (2004, 129–30).

view about the persistence of a subject in generation. When Aristotle later reflected on the issue in *GC* he came to the view that it does not require an enduring subject. Instead whether or not a change meets the Persistence Condition serves as the dividing line between qualified and unqualified becoming.

There are two ways that a defender of the standard reading might respond to this. One strategy is to lean heavily on Aristotle's statement in T2 that it is alteration 'when the *perceptible* subject remains but changes in its own properties', whereas in generation 'nothing *perceptible* remains as its subject'. One might claim that this leaves open the possibility that in generation something *imperceptible* survives the change, namely, prime matter.[10] This would bring the *GC* model in line with the standard reading of *Physics* I 7, since (on this reading) prime matter remains as a continuant of the change in generation. However, there are several reasons for resisting this interpretation.

First, none of the other core passages in which Aristotle draws the distinction between generation and alteration make any reference to the perceptibility (or otherwise) of the subject. It is not part of the original characterisation of the dispute at *GC* 314b1–6. Nor is it mentioned in T1 or T3. T2 is alone in referring to the perceptibility of the subject in this context. This suggests that Aristotle did not see perceptibility as crucial to the distinction he wants to make between substantial generation and non-substantial changes but only whether the underlying subject persists or not. More importantly, this interpretation depends on an argument from silence. There is no mention in *GC* of any imperceptible continuant of generation like prime matter. And yet, if Aristotle did mean to say that some imperceptible subject persists in generation, we should expect him to make that crucial point right there in T2, especially in light of the way he originally set up the contrast between monism and pluralism in *GC* I 1 (where the perceptibility of the subject is not at issue).[11] So at best T2 leaves room for something like prime matter to function as a persistent subject in generation; it does not explicitly propose that idea.

But there are reasons for thinking that this is not what Aristotle has in mind in T2. When he says we consider it generation 'especially when the change proceeds from an imperceptible something to something perceptible', his example of an imperceptible subject is not some metaphysical

[10] Philoponus (*in GC*, 66,12–17), Krizan (2013, 306–7). See also Robinson (1974). This alternative is considered and rejected by Charlton (1970, 135), Gill (1989, 47–53), and Broadie (2004, e.g. 127–9).

[11] For more discussion of this point see Gill (1989, 42–53) and Broadie (2004).

postulate like prime matter but ordinary air ('for air is pretty imperceptible').[12] And this is likely just a nod to the commonsense account of generation, which relies on the criterion of perception. More importantly, Aristotle goes on to reject the relative perceptibility of the subject as a true standard for distinguishing generation from alteration at *GC* 318b18–28. Clearly, when Aristotle refers to imperceptibility in these contexts, neither he nor common sense has in mind any sort of metaphysical postulate like prime matter. In light of this, I find it difficult to believe that by saying nothing *perceptible* remains as a subject Aristotle means to suggest that prime matter is the *imperceptible* subject.

Finally, the focus on perceptible versus imperceptible subjects is a red herring. For it makes no difference to how Aristotle actually distinguishes the two sorts of changes in *GC* I. This is clear from the way he divides up the opposing camps in *GC* I 1–2. On the one hand, the monists are said to have reduced generation to alteration because they held that 'in all coming-to-be there is a single element that serves as a persistent substratum', which Aristotle calls alteration ('*we* call this sort of thing being altered'). Presumably Aristotle means to include in this camp monists like Anaximenes and Anaximander whose own theories include something imperceptible as the enduring subject (air and the *apeiron*, respectively). If so, then he must be willing to treat some change as an alteration even when some imperceptible subject persists. On the other hand, Aristotle treats atomism as the sort of theory that distinguishes generation from alteration. Since atoms (which are also imperceptible) remain through the change, it cannot be the imperceptibility of what endures that determines whether or not generation is reducible to alteration. All that matters, it seems, is whether or not the underlying subject of generation persists as numerically the same subject throughout regardless of its perceptibility.[13]

Another way to defend the standard reading would be to reinterpret what Aristotle means in *GC* when he says that the subject of generation is 'completely transformed' (*holon metaballê*) in the process of becoming a new substance. For example, when Aristotle says that the subject does not remain but is replaced by a new subject, what he could mean is that it does not remain *what it was* at the start of the change because it lost the

[12] Cf. Brunschwig (2004, 41 n. 39).

[13] I have also argued in Chapter 1 that Aristotle has good reasons for rejecting prime matter as the proper subject of generation (*GC* I 3). For it entails either that generation proceeds from nothing at all, or that 'what is not a determinate "this" (*tode ti*)' is nevertheless capable of independent existence, or that properties could exist in their own right independently of substances – all consequences that Aristotle finds unacceptable.

form that determined its original identity. This leaves room for saying that the subject does remain qua matter. On this reading matter would still serve as a continuant even in cases of substantial generation.[14]

However, when Aristotle says that in cases of substantial generation the subject does not remain, he never once qualifies that statement by saying it remains qua matter. On the contrary, he explicitly denies that. In T1 Aristotle says that one part of the subject corresponds to its form and the other to its matter and that generation occurs when there is a change 'in these' (not 'in one of these'). The natural antecedent for the plural *toutois* is the form *and* matter (not just the form).[15] This is why Aristotle says that in generation one subject '*wholly* changes' into another (*metaballê ek toude eis tode holon*).[16] This is also the point of the contrast in T2 between the coming-to-be of a statue, where the bronze survives the change from spherical to angular, with genuine cases of substantial generation, where the subject is transformed entirely. In the former, but not the latter, the subject qua matter remains as a continuant. If Aristotle meant to say that the subject of generation is only transformed with respect to its form while its underlying matter persists, then saying 'one component of the subject corresponds to the form and the other the matter' and that 'generation and destruction occur when there is a transformation in these' is a very misleading way of making that point.

Given that Aristotle does not explicitly argue in *GC* that matter persists through the generation of a substance, we might reasonably wonder why someone would attribute that view to him there. One obvious motivation is to make that account fit with what he says in *Physics* I 7. But we have already seen that the evidence from *Physics* I 7 is not conclusive enough to force that interpretation on us.

Another possibility is that Aristotle needs to hold that matter persists through the change in order to meet some demand on his theory. For example, Gill argues that Aristotle makes matter an enduring subject in order to preserve what she calls the 'horizontal unity' of a substance (its continuity over time).[17] But in *GC* Aristotle does not raise the question

[14] Gill (1989, e.g. 6–7) adopts this kind of strategy. The reading she develops is quite a bit more nuanced than the standard interpretation (see note 23 below). However, I think it is ultimately unsuccessful as a reading of *GC*, because it fails to treat substantial generation as *sui generis* but assimilates it to mixture (*mixis*). See Section 3.4 below.

[15] Williams (1982, 79). [16] Joachim (1926, 88).

[17] Gill (1989, e.g. 6–8). The demand for horizontal unity is not just a demand that there be something to ground the continuity of the individual through its own existence (from the cradle to the grave) but also for something to secure continuity with the pre-existing entity from which it came to be. This seems to be why she connects the demand for horizontal unity with Aristotle's desire to avoid

of persistence in the context of a worry about the continuity of a substance over time but in the context of a debate over whether substantial generation is reducible to alteration. And, as we have seen, Aristotle's own position in that debate turns on denying that generation involves an enduring subject (314b1–6). Another common view is that Aristotle introduced the Persistence Condition to avoid treating substantial generation as "sheer replacement".[18] According to this argument if nothing persisted through the change, then Aristotle would have to admit that one entity is destroyed into nothing and another entity emerges from nothing, which violates Eleatic principles that Aristotle supposedly accepts. But again, there is no sign in *GC* that Aristotle was concerned with sheer replacement let alone that he posited matter as a continuant in order to avoid that. Aristotle is certainly concerned to avoid the idea of generation *ex nihilo* in *GC* I 3. But his worry there is about how to characterise the subject of generation given his commitment to the idea that *if* something comes to be in the unqualified sense then it must proceed from what is not in the unqualified sense (see Chapter 1). At no point does Aristotle raise the problem of how to reconcile his claim that in generation the subject does not survive with concerns about sheer replacement. I think that as long as we do not go beyond the text of *Generation and Corruption*, there is no good reason for ascribing to *GC* the view that in substantial generation matter must remain as a continuant.

According to the interpretation defended in this chapter, the *GC* model characterises substantial generation in terms of what we might call a "radical transformation" in the subject that undergoes the change: in the process of becoming a new substance the subject is changed as a whole, which Aristotle glosses in T1 as a transformation of its matter and form together. This allows him to treat substantial generation as *sui generis*. In cases of substantial generation something genuinely new comes into being that was not there before whereas in non-substantial changes (growth, alteration, locomotion) the subject remains one and the same and simply changes with respect to its incidental properties. I have also argued that this doctrine is compatible with *Physics* I 7 if we adopt the more restrictive reading of that chapter. One of the things I tried to emphasise in Chapter 2 is that, although *Physics* I 7 is explicit that the subject survives in those cases where a thing comes to be in the qualified sense, it does not reflect on, let

the Eleatic objection against "sheer replacement" (see below). On her account, this demand is only met by making matter the continuant so that 'the matter from which a composite is generated remains "present in" the composite' (in some nuanced sense) (Gill 1989, 109).

[18] Again Gill (1989, 7, 241), but also Code (1995).

alone defend, the claim that this also holds for substantial generation. This is not to say that in *Physics* I 7 Aristotle explicitly denies that substantial generation involves a persisting subject; he does not. Instead I have suggested that the model of generation on offer there is simply undeveloped with respect to the Persistence Condition. As Broadie (2004, 129–30) puts it, 'although in *Physics* I. 7 Aristotle was definite that the coming to be of substance is the coming to be of something in a different category from the others, he did not dwell on what difference this implies for the logical structures of the corresponding types of change'. Thus, while there is some sort of development between the *Physics* and *GC* when it comes to substantial generation, this does not mean Aristotle changed his mind about anything. For the *Physics* does not defend any substantive thesis about the persistence of a subject in generation which is then later rejected in *GC*. And so we do not have to see the development as a change in philosophical position. Instead the *GC* account represents a refinement of Aristotle's views that results from having come to reflect more carefully on the differences between qualified and unqualified becoming. This would be true even if in *Physics* I Aristotle simply took it for granted that persistence is among those 'common features' of all coming-to-be (189b31) without actually reflecting on the issue.

This way of understanding the difference between *Physics* and *GC* (as a shift from an unreflective to a more reflective view about generation) can be illustrated by considering what Aristotle says about the case of a statue coming to be from bronze. Both *Physics* and *GC* agree that in this case the subject (the bronze) survives the change. Where they disagree is over the ontological status of that change. Defenders of the standard reading take Aristotle's use of the statue example at *Physics* 190a25–6 to be making a point about substantial generation, where Aristotle says that the bronze survives.[19] And the later passage at 190b5–9 certainly invites this. There Aristotle includes statues among those things 'that come to be without qualification', and he has just said that only substances come to be in the unqualified sense.[20] Yet, at *GC* 319b8–21 Aristotle explicitly describes statue-making as an *alteration* of the bronze, which he contrasts with

[19] E.g. Code (1976, 357), Ferejohn (personal communication). The passage reads: 'We speak of "becoming that from this" instead of "this becoming that" more in the case of what does not survive the change – "becoming musical from unmusical", not "from man" – but we sometimes use the latter form of expression *even of what survives*; we speak of a statue coming to be from bronze, not of the bronze becoming a statue.' (190a21–5, Hardie and Gaye translation)

[20] Henry (2015a, 159) suggests that *Physics* I 7 does not mean to count statue-making as a case of substantial generation. But *Physics* 190b5–9 appears to do that after all.

coming to be in the strict and unqualified sense on the grounds that in the former but not the latter the subject survives. This conflict extends beyond artefacts, since *Physics* 190b5–9 includes all sorts of changes that *GC* explicitly distinguishes from substantial generation on the same grounds, including growth and alteration.[21]

On the standard reading this disagreement reflects a fundamental change in position. On the restrictive reading, however, this does not mean Aristotle changed his mind between *Physics* I and *GC* I. For unlike *GC* Aristotle's remarks in the *Physics* do not stem from a carefully developed position about how substantial generation differs from changes in the non-substantial categories. When *GC* 319b8–21 contrasts the production of a statue (where the bronze survives and is altered from spherical to angular) with genuine cases of substantial generation (where the subject is 'wholly transformed') it comes in the context of a sustained inquiry concerning the differences between qualified and unqualified becoming. By contrast in *Physics* I 7, while Aristotle is clear that only substances come into being without qualification, he does not reflect on what exactly counts as substantial generation and how it differs from changes in the non-substantial categories. And so *Physics* 190b5–9 simply lumps together all sorts of changes that *GC* is careful to separate off from substantial generation by reference to the Persistence Condition. Thus, while Aristotle's views differ between the two works this difference does not represent a change in philosophical commitments. For *Physics* I 7 does not represent Aristotle's carefully developed views on the question of whether or not substantial generation involves a persistent subject.[22]

[21] Code (1976, 365) tries to explain this away by saying that in *GC* Aristotle thinks we can describe the same change as both an alteration and a substantial generation. And so one can treat these changes as alterations in one context and substantial generations in another. (Code makes this point explicitly with regards to statue-making, though he wrongly cites *GC* 318b13–14 as evidence for that.) However, as we have seen, part of the aim of *GC* I 1–4 is to establish substantial generation as a distinct change in its own right by showing how it is something separate from alteration (314a1–6). There is no indication that Aristotle thinks this distinction is relative to a given perspective. If he wanted to make that point, we would have expected him to do so in *GC* I 4 in the context of the worry about elemental transformations. As we have seen, elemental transformation always involves some persistent item, which might make it look like an alteration. Aristotle does not answer this by suggesting that the same change can be viewed as an alteration from one perspective and substantial generation from another perspective. Rather, he argues that elemental transformation still counts as substantial generation *and not an alteration* on the grounds that it only counts as an alteration when *the subject* persists, and qualities do not count as proper subjects.

[22] For this reason, the argument of this chapter doesn't turn on the relative dating of *Physics* and *GC*. All that is important is that in *GC* I 1–4, Aristotle carefully reflects on how substantial generation differs from other sorts of changes. And so, we should take his views in *GC* as having authority over *Physics* I 7 (which does not reflect on their difference).

3.4 Generation versus Mixture

As we have seen, Aristotle's main project in *GC* I 1–4 is to show that substantial generation is *sui generis* and that it is not reducible to a form of alteration (coming-to-be in the qualified sense). He does this by arguing that in cases of alteration the subject persists and changes merely in its incidental properties (e.g. 'the body, although enduring as the same body, is now healthy and now ill, and the bronze is now spherical and now angular, and yet remains the same bronze'), whereas in generation the subject does *not* persist but is completely transformed in the process of becoming a new substance (see T1 to T3 above). This was precisely how the debate was originally framed in *GC* I 1:

> It is necessary for those who construct all things out of a single element to say that coming-to-be and passing-away are alteration. For the underlying subject [sc. that persisting element] always remains one and the same, and we call that sort of change "being altered". (*GC* 314b1–6)

One might respond by saying that the kind of theory Aristotle is targeting in this passage treats matter as something that survives *intact* and remains *actually* present in the finished product. Thales, for example, is supposed to have held that water persists as an underlying subject because of the ubiquitous presence of moisture within natural bodies:

> Thales, the founder of this school of philosophy, says the principle is water (for which reason he declared that the earth rests on water), getting the notion perhaps from seeing that the nutriment of all things is moist, and that heat itself is generated from the moist and kept alive by it (and that from which they come to be is a principle of all things). He got his notion from this fact, and from the fact that the seeds of all things have a moist nature, and that water is the origin of the nature of moist things. (*Metaph.* I 3, 983b19–27, Ross translation)

Perhaps Aristotle simply means to contrast this sort of view with one where the subject of generation is *actually* destroyed and survives in the generated compound only as a *potential* constituent.[23]

[23] Gill (1989) defends this more nuanced view. See also Carraro (2017, 279 n. 11). Gill cashes this out by saying that what persists in generation is only a set of dispositional properties that characterise both the pre-existing matter and the matter of the product that comes to be from it (e.g. the wooden box retains many of the dispositional properties that characterised the separately existing wood from which it was constructed). Gill calls this surviving complex of dispositional properties 'generic matter', which is not a definite subject (*tode ti*) and has no separate existence of its own. Rather, it is something 'potential and indefinite, like a genus, and substantial form differentiates the generic matter into some definite thing' (Gill unpublished; cf. Gill 1989, 163). In Gill (unpublished) she argues that Aristotle endorsed the *Physics* I 7 model (traditionally understood) in *Metaphysics* VII to

Aristotle certainly does recognise changes where the pre-existing matter survives in the product potentially. It is the essence of what he calls combination or mixture (*mixis*):

> Since, however, some things exist in potentiality while others exist in actuality, the constituents [of a compound] can *be* in a sense and yet *not-be*. The compound may be *actually* different from the constituents from which it has resulted; nevertheless, each of them may still remain *potentially* what it was before they were combined, and both of them may survive without perishing. For this was the difficulty that emerged in the previous argument, and it is clear that the combining constituents not only coalesce, having previously existed in separation, but can also again be separated out from the compound. The constituents, therefore, neither actually persist (as body and white persist) nor are they destroyed (either one or both of them), for the potentiality (*hê dunamis*) is preserved. (*GC* I 10, 327b22–31, translated after Joachim)

Gill (1989, 147–9) takes this passage as a model for understanding how matter serves as a continuant in substantial generation. In a mixture,[24] the pre-existing matter is not completely destroyed in the process. And yet it does not survive without qualification (*haplôs*: 328b14–19). Instead, the matter survives only potentially and is recoverable from that product through a process of chemical analysis.

But even this more nuanced version of the standard reading fails to treat substantial generation as *sui generis*. For Aristotle is explicit in *GC* I 10 that *mixis* and *genesis* are two distinct processes and that the failure to keep them apart is at the root of those puzzles that purport to show that *mixis* is impossible:

> According to some thinkers it is impossible for one thing to be combined (*mixthênai*) with another. On the one hand, they argue that if the combined constituents continue to exist and are not altered, then they are no more combined now than they were before but are in the same condition. On the other hand, if one has been destroyed, then the constituents have not been combined; on the contrary, one constituent exists and the other does not exist, whereas mixture demands uniformity of condition in them both. And on the same principle, even if both the mixed constituents have been

VIII 1–5 but not in *Metaphysics* VIII 6 and IX. In the latter, the pre-existing matter from which a substance comes to be does not remain in the product (except potentially) but is *transformed into* that substance.

24 What I am translating as 'mixture' or 'combination' is the change Aristotle calls *mixis*, in which the material ingredients survive only potentially in the product. By contrast, Gill translates *mixis* as 'compound' or 'chemical combination', while 'mixture' is used for *sunthesis* where the original ingredients are preserved in the product intact.

> destroyed as a result of their coalescence, *they* cannot be what is combined since they have no existence at all [sc. in the compound]. What we have in this account it seems is a demand for what it is precisely that differentiates mixture from generation and destruction, and for the precise distinction between what is mixable and what is generable and destructible. For it is clear that mixture, if it exists, must be different from those processes. And so the difficulties raised by the account would be resolved as soon as these distinctions are made clear. (327a34–b10, translated after Joachim)

Aristotle does not dwell on the distinction between mixture and generation in *GC* I 10. All he says in defence of this claim is that we tend not to speak of generation in terms of mixture. For example, when wood is burned, we do not say that the wood is mixed with the fire; rather, we say the wood is destroyed and fire comes into being (327b10–13). But that is sufficient for his point here because he has already established in *GC* I 1–4 what generation is and how it is differentiated from non-substantial changes. And so, by making clear what mixture is and what things are mixable, *GC* I 10 puts us in position to see how this sort of change differs from substantial generation. Mixture still counts as a form of alteration (broadly construed) because it involves a persistent subject that remains present in the resulting compound. It differs from normal alterations because, whereas in those other cases the subject survives intact to become an actual constituent of the finished product (see T3 above), with mixtures the original materials persist only in potentiality. Since Aristotle is clear that mixture is distinct from substantial generation, it stands to reason that when a new substance comes to be in the unqualified sense there is an even more thorough and complete destruction of the pre-existing matter than this insofar as it does not survive the change *even potentially*.[25]

3.5 Two Puzzles about Persistence

The restrictive reading defended in the last two chapters has several virtues over the orthodox interpretation (which takes the Persistence Condition to apply even in cases of substantial generation). Not only does it avoid the consequence of having to say that Aristotle changed his mind about

[25] Frede (2004, 290) emphasises the odd status of *mixis* in Aristotle's ontology of change: 'It is neither the result of simple substantial change in the sense of generation proper, nor a straightforward case of alteration, nor a change in quantity, nor is it locomotion.' See also Frey (2007, 187). But it is not that odd if we consider that *mixis* must still count as a form of coming-to-be in the qualified sense (i.e. an "alteration" broadly construed). For the dichotomy between qualified and unqualified becoming is exhaustive, and mixture does not count as coming-to-be in the qualified sense.

persistence between *Physics* and *GC*, it also avoids two awkward puzzles that arise from the standard view: if Aristotle did think that the pre-existing matter from which a new substance comes to be persists in the finished product, then it is difficult to see how his two prime examples of substantial generation – biological generation and elemental transformation – conform to his proposed model. For in both cases it is puzzling what this supposed continuant could be.

3.5.1 *A Puzzle about Biological Generation*

Suppose we accept the standard reading of *Physics* I 7 that the pre-existing matter persists through a substantial change to become a constituent of the resulting compound. While this doctrine certainly makes sense for artefacts like brick houses and bronze statues where the pre-existing matter clearly survives in the finished product, it seems fairly implausible in the case of organisms. For what the *Physics* account specifies as the subject of biological generation is the seed (190b4–5). But this matter does not survive in the organism (even potentially). Instead, the matter that serves as the underlying subject of the mature form is some more proximate matter, namely, a body composed of functioning organs (the '*organikon* body', *DA* 412a28–b6). Freeland puts the puzzle this way:

> [I]t seems impossible to point to something that serves as both matter "for" and matter "of" a living creature,[26] as bronze does for statues. In *Physics* I.7 Aristotle tells us that in any change some underlying thing or substratum persists, and it is easy to see that the bronze persists when a statue is molded, the gold continues in the bangle, etc. But what exists before a person, duck, or oak tree comes to be is an embryo, egg, or acorn (cf. *Ph.* I.7, 190b3–5) – and these surely do not remain in the finished creature as their matter.[27]

In the language of Chapter 2, the puzzle about biological generation arises because the seed satisfies the Subject Condition but not the Persistence Condition. For living things come into being from seed, but this clearly

26 The distinction between "matter for" and "matter of" derives from Code (1976), which I assume refers to the distinction between the pre-existing matter out of which a new individual comes to be (e.g. seed) and the constitutive matter out of which its body is composed (e.g. bones and tissues). See note 27 below.

27 Freeland (1987, 392). See also Gill (1989, 161–2). Code (1987, 55–6) puts the puzzle this way: 'The idea that the genesis of a hylomorphic composite requires the pre-existence of both the matter and form raises the following difficulty. In the case of the genesis of Bucephalus the matter which pre-existed is some menstrual fluid in his mother's womb, whereas his current matter is not a quantity of menstrual fluid, but rather a highly articulated structure – an organic body potentially having life in it.'

does not survive the change (even potentially) to become a constituent of the mature organism.

Freeland argues that the biological works actually vindicate the orthodox reading of *Physics* I 7 in a way that resolves the tension between these two conditions. For in several places Aristotle identifies blood as both the pre-existing matter out of which an animal ultimately comes to be (insofar as menstrual fluid is a form of blood) and as the proximate matter for the parts of the mature animal (insofar as blood is the matter for the parts: e.g. *PA* 651a13–15, 668a7–8, *GA* I 18–22).[28]

But Aristotle's mature account of animal generation does not vindicate the standard reading. Even if blood is a good candidate for the matter that persists through biological generation, the *GA* never raises a puzzle about what persists through the change let alone suggest that blood is what satisfies some demand for a persistent substratum. When taken in context neither the claim that menstrual blood is the matter for generation nor the claim that nutritive blood is the matter for the parts of the mature animal is introduced as a way to satisfy PC. The closest Aristotle comes to making this claim in the biological works is at *GA* 740b35–6, where he says that the matter from which the mature animal grows and the matter out of which it is first constituted are the same. The main thesis of this passage is that the same power of soul is responsible both for constructing the parts of the organism at the very beginning and for growing those parts later on. The inference appears to depend on the *De anima* principle that activities are individuated by their objects and capacities by their activities (*DA* II 4, 415a14–22). Since the matter of these two activities is the same (both operate on blood), it follows that their respective capacities must be the same as well. That argument in no way depends on the claim that this matter also persists through the change.

Nor should this thesis be taken to entail that blood is the persistent substratum of biological generation. Aristotle holds that blood is both the matter out of which the organism grows (coming-to-be in the category of quantity) and the matter out of which it is first constituted (coming-to-be in the unqualified sense). Nothing about this entails that blood is the *terminus a quo* and *terminus ad quem* of generation (as Freeland's account assumes). It simply a thesis about the *terminus a quo*: it says that these two

[28] Even if this were Aristotle's view in the biological works, it clearly conflicts with the *GC* doctrine. As we saw, Aristotle characterises the monists as having reduced generation to the alteration of a persisting subject (314b1–3). If Freeland is right, then it isn't clear why Aristotle would be innocent of the same charge that he himself reduces animal generation to the alteration of some persistent subject (e.g. that it is just a modification of some underlying blood).

ontologically distinct kinds of change *begin from* the same kind of matter.[29]

In Chapter 3 I argued that when Aristotle came to reflect carefully on the differences between substantial generation and changes in the non-substantial categories (*GC* I 1–4), he identified the (non-)persistence of the subject as the key distinction. According to the *GC* model, the matter that serves as the subject from which substantial generation begins does not persist but is completely transformed in the process of becoming a new substance. I see no reason to think the biological works are saying anything different from that. For example, the *GA* relies on a theory of generative "residues" that emphasises the different kinds of materials that are involved in constructing the new animal. In *GA* II 6, for example, he divides the matter for generation into (i) spermatic (*spermatikon*) and nutritive (*threptikon*) matter, which 'provides both the whole and the parts with being', and (ii) supplemental (*epiktêton*) and growth-promoting (*auxêtikon*) matter, which contributes to the bulk of those parts after they are formed (744b32–745a4; cf. 776a31–b3). The second category includes milk (in live-bearing animals) and the additional food the child consumes after birth. The first includes not only menstrual blood, which Aristotle calls "first matter" (*tên prôtên hulên: GA* I 20, 729a32–3) insofar as it is used to constitute the original embryo, but also a separate supply of uterine blood that is acquired through the umbilical cord.[30] During development these materials are transformed into other kinds of residues that form the raw materials for the uniform parts (e.g. *GA* 744b12–745a20), which in turn become the matter for various instrumental parts (*PA* 646a13–b28) that together make up the living body that serves as the proximate matter for the soul of the mature organism (e.g. *DA* 412a18–20). Aristotle gives no indication that blood (or any other more fundamental kind of matter) endures through all of these complex transformations as some kind of persisting substratum. It is perfectly consistent with this picture to say that, although those materials can all trace their causal history to the blood that comes from the mother, that original blood does not remain but is entirely transformed in the process of becoming the new substance.

Of course, this is not definitive enough to conclude that Aristotle actually rejected the Persistence Condition in the *GA*. Nothing about this picture shows that blood (or even prime matter) does not persist

[29] For a further critique of Freeland's interpretation see Connell (2001, 309–10).
[30] See Chapter 1, note 30.

through the cascading series of transformations that makes up the gradual transition of the embryo from a prior state of potentiality to a more mature state of completion. However, the fact remains that the *GA* never identifies blood (or any other material) as the thing that satisfies some requirement for a persistent subject. At best blood constitutes a possible answer to a question that Aristotle never poses in his mature writings on living things. The puzzle about organisms seems obvious enough that if Aristotle did think that substantial generation must satisfy PC, then we should have expected him to address that somewhere in *GA* since living things are paradigm natural substances and biological generation is substantial generation par excellence. And yet, among the many *aporiai* that Aristotle investigates in the *GA*, he never once raises a puzzle about what, if anything, plays the role of the persisting substratum. Questions about material continuants are simply not on Aristotle's radar in *GA*. That would be a striking oversight, if the standard reading were correct.

The virtue of the restrictive reading, then, is that the puzzle about organisms disappears. There is nothing that serves as both the pre-existing matter out of which a new organism comes to be and also survives to become the constitutive matter out of which its mature body is composed as bronze does for statues. But this is not a problem for the restrictive reading. For Aristotle does not demand (at least in his more reflective writings on substantial generation) that the pre-existing matter survives the change to become a constituent of the substance that comes to be from it. And so there is nothing puzzling about the fact that 'what exists before a person, duck, or oak tree comes to be' (embryo, egg, acorn) is not the same as the matter that constitutes the finished creatures.

3.5.2 A Puzzle about Elemental Transformation

The standard reading also generates a puzzle about the mutual transformation of the elements into one another. Elemental transformation is supposed to count as substantial generation. But it is not clear what, if anything, serves as the persistent matter of the change. If nothing does play that role (the argument goes), then elemental transformation is not explicable in terms of matter, form, and privation as the *Physics* I 7 model suggests.[31]

[31] For the puzzle about the elements see Krizan (2013) and references therein.

Defenders of the standard reading might reach for one of two things to serve as the enduring matter of elemental change. One is prime matter.[32] I have offered several reasons for thinking this was not Aristotle's view, and I will not pursue this option further here. Another is to count one of the surviving contraries as the matter that serves as the enduring subject of the change.[33] For example, when air (hot-moist) comes to be from fire (hot-dry) the hot remains. However, as we have already seen, Aristotle rejects this possibility in *GC* I 4. Aristotle's worry there is that if we treat the hot (the persisting quality) as an enduring subject, then elemental transformation will turn out to be a form of alteration rather than substantial generation (319b21–4). For, according the *GC* model, a change counts as an alteration whenever *the subject* remains and exchanges one property for another. Aristotle avoids this problem by denying that the persisting quality counts as an enduring *subject*. This is why he says at the end of *GC* I 4 that a change qualifies as generation when nothing persists 'of which the resultant is a property' (319b34–320a2). In cases of elemental transformation, the resultant (e.g. the moist) is *not a property* of the continuant (e.g. the hot), since qualities cannot serve as subjects of predication. And so, what persists is not a proper subject (*hupokeimenon*).[34] If this is right, then Aristotle does not think the primary contraries involved in elemental transformation play the role of persistent matter for the change.

The interpretation defended in the last two chapters allows us to see how the transformation of the elements into one another can be accommodated within Aristotle's basic hylomorphic model of generation in *Physics* I 7. On that interpretation, Aristotle only requires that substantial generation proceeds from some pre-existing thing that meets the conditions for being an ultimate subject of predication (what I called the Subject Condition). Aristotle is not committed to the further claim that this pre-existing thing must also persist through the change to become a constituent of the resulting substance (substantial generation does not have to meet the Persistence Condition). In the paradigm cases of biological generation, the

32 Robinson (1974), Krizan (2013), cf. Charles (2004). 33 See Krizan (2013, §3.2).

34 See Broadie (2004, 130–3). This rules out solution (II) discussed by Krizan (2013, 300). Frey offers a different argument for the same conclusion. He denies that in elemental transformations the quality persists in the product in any robust sense: 'for the hot to persist is not for there to be some autonomously identifiable element that at one time is paired with the dry and at a later time paired with the wet'; instead, he argues, the original quality persists only in the sense that (e.g.) the hot of fire and the hot of air 'manifest themselves similarly with respect to their active changes' (Frey 2007, 183).

Subject Condition is satisfied by the seed from which the new organism develops (*Ph.* 190b4–5), e.g. the acorn out of which the new oak tree grows.[35] In cases of elemental transformation, it is satisfied by the element that serves as the *terminus a quo* of the change, which is destroyed in the process of becoming a new element (as *GC* I 1–4 requires of generation). For example, when air comes to be from fire, fire itself serves as the pre-existing subject from which the change proceeds. This change counts as substantial generation precisely because the fire (the original subject) passes away and air (the new subject) comes to be. This is enough to show that elemental transformation conforms to the bare bones model of *Physics* I 7, which identifies the principles of natural generation with matter, form, and privation. In our example, fire is the subject from which the change proceeds, moist is the form that is acquired as a result of the change, and dry is the privation that is lost in the change.[36] However, as we shall see in Chapter 4, elemental transformation does not conform to the richer model presented in *GC* II 9. The reason, I shall argue, is that the teleological dimensions introduced into the model do not apply to cases where one element changes into another or the phenomena that those changes underwrite (e.g. seasonal rainfall): while the resultant quality (e.g. moist) serves as the form acquired as a result of the change, it does not play the role of *the end* or *goal* for the sake of which that change takes place; nor does the process involve any goal-directed efficient cause that directs the process towards that form. Both of these are essential aspects of what I call the 'extended' hylomorphic model of substantial generation that Aristotle develops in *GC* II 9.

[35] There is a lingering question here, to which I shall return, about what exactly is "the seed" that constitutes the subject of animal generation (see Chapter 5, Section 5.6). While the *GA* appears to identify this with menstrual blood, the doctrine of potential being in *Metaphysics* IX 7 suggests that it should be the embryo once its rudimentary heart has been differentiated out of it and it begins to process nutriment from the mother. I attempt to reconcile these two positions in Chapter 5 by drawing on the distinction between embryogenesis (whose product is the embryo) and morphogenesis (whose product is the fully-articulated offspring). To anticipate my solution, menstrual blood is the subject from which the first proceeds while the embryo that results from that change becomes the new subject for the second.

[36] I follow Krizan (2013, 305) in taking the form in an elemental transformation (in the *Physics* I 5–7 sense) to be the contrary that comes to be as a result of the change.

CHAPTER 4

The Extended Hylomorphic Model: GC *II 9*

As we have seen *Physics* I identifies matter and form as 'the causes and principles' from which naturally generated substances first come into being (e.g. 190b10–23; cf. *Metaph.* VII 8). On this model, matter plays the role of the subject from which the change proceeds while form is the positive state acquired as a result of the change. When Aristotle returns to the principles of natural generation in *GC* II 9, he goes to great lengths to show that these two principles alone are not sufficient to account for the change. Instead, we must posit a third principle over and above those two:

> The principles are equal in number to, and identical in kind with, those in the sphere of the eternal and primary things. For there is one in the sense of matter and another in the sense of form. But there must be a third one present over and above these (*eti prosuparchein*); for these are not sufficient to bring things into being any more than they are to account for the primary things. (335a28–32)

While Aristotle does not say explicitly what this third principle is, the context makes it clear that he has in mind a source of motion. He defends this extension of the hylomorphic model by arguing that previous attempts to explain natural generation in terms of matter and form alone were unsuccessful precisely because they did not avail themselves of this kind of cause. Before turning to Aristotle's arguments for this, I want to make some preliminary remarks about how *GC* II 9 extends the basic model examined in the previous two chapters.

4.1 The Principles of Natural Generation Revisited

In addition to extending the *Physics* I 7 model to include a source of motion, *GC* II 9 also refines the way that matter and form serve as principles of generation (and destruction) (335a32–b6). Matter is a principle insofar as it makes things capable of being and not being, while form is a principle in the sense of the end for the sake of which the change takes place. Neither of these ideas plays a significant role in *Physics* I 7's presentation of the model.

4.1.1 Matter

Let me start with the refined conception of matter. Aristotle first explains how matter is a principle and cause of generation and destruction by associating it with a certain kind of capacity:

> One principle is a cause of generable things as matter insofar as this is what is capable of being and not being (*to dunaton einai kai mê einai*). For some things exist from necessity (e.g. eternal things) while others necessarily do not exist. Of these, because it is not possible for them to be otherwise than in their necessary state, the first are incapable of not existing while the latter are incapable of existing. But other things are capable of both existing and not existing, which are precisely those that are generable and destructible. For this type of thing exists at one time and does not exist at another. Therefore, it is necessary for generation and destruction to occur among those things that are capable of existing and not existing. This is why in one sense matter is a cause for the things that come to be. (*GC* 335a32–b6)

I will examine this argument in more detail in Chapter 9 where I consider the place of sublunary generation in Aristotle's broader cosmology. The relevant point to take away from this passage is that being composed of matter helps explains why natural substances come to be and pass away in the first place insofar as matter is the seat of a thing's capacity to be and not be.[1]

As I understand *GC* II 9, when Aristotle says that matter is a 'cause' of natural generation 'insofar as this is what is capable of being and not being'

[1] Aristotle makes the same point at *Metaphysics* VII 7 where he says that all generable things contain matter: 'for each of them is capable of existing and not existing, and this is the matter in each of them' (1032a20–2). I take *touto* (a21) to refer back to the capability (*dunaton*) that each generated substance has for being and not-being, which is the intrinsic source of its generability and destructability (cf. *Metaphysics* VI 1, 1026a2–3). This capacity concept of matter forms part of a whole family of ideas, including the idea of matter as potential substance and matter as genus. I will not pursue these complex connections in any detail here. In *Metaphysics* XII Aristotle says that there are different kinds of matter that make things susceptible to different kinds of change: 'All things that change have matter but their matter is different. Even eternal things that are not generable but are movable in place have matter, not generable matter but matter for moving from place to place' (1069b25–6). By "generable" matter I presume Aristotle has in mind the capacity to be and not be that is the focus of *GC* 335a32–b6 (the kind that is a principle of generation and destruction). The heavenly spheres lack this sort of matter, which is why they are not subject to these changes. However, they do contain "locomotive" matter ('matter for moving from place to place'), which is a capacity to be and not be *at a place*. Aristotle may have in mind here the so-called fifth element whose nature is to move in a circle.

(*einai kai mê einai*), he means being and not being *a substance*.[2] Christopher Shields (2007, 141) argues that Aristotle's concept of substantial being is (as he puts it) sortally determinate: 'By this I mean that substance is not merely the cause of being in the sense of being responsible for the existence of a thing, but also the cause of its *being the F kind of thing it is*.' Aristotle expresses this point in *DA* II 4 by saying that 'substantial being (*ousia*) is the cause of being (*to gar aition tou einai*) for all things', for example, to be for an organism is to be alive (or to be a living thing, 415b12–14). Aristotle seems to apply this principle all the way down to the level of atomic species (e.g. humans, lions, cranes). In *PA* I 1 he makes it clear that these species constitute the true substantial beings (*ousiai*) of the living world.[3] This suggests that the concept of generable matter should be taken as relative to the kind of substance that is coming to be. For example, the generable matter of a horse is a capacity to be and not be a horse, which is different from the generable matter of a human which is a capacity to be and not be a human. If this is right, then every atomic species will have its own generable matter which is a capacity to be and not be that particular kind of substance.

There thus seem to be two ways that Aristotle conceptualises the matter of generation. In *Physics* I 7 matter is primarily associated with the subject from which the change proceeds. Here Aristotle appears to be thinking of some kind of determinate material stuff like bronze or wood or seed. He calls this 'the countable matter (*hê hulê arithmêtê*)' since 'it is more like a determinate individual' (*tode ti*).[4] In *GC* II 9 matter comes to refer to that feature of the subject that is the seat of its capacity to be and not be a substance of a particular kind, which is supposed to explain why certain kinds of things are capable of generation and destruction. For example, *Physics* 190a3–5 identifies the seed as the proper subject from which plants and animals first come to be, which makes it the matter of

[2] With the following compare the discussion of female matter in Connell (2016). See also Ebrey (2018). As Joachim (1926, 246) notes, the capacity concept of matter in *GC* II 9 is 'a restatement in somewhat modified terms' of the doctrine implied by *GC* I 3 (see Chapter 1).

[3] *PA* 639a15–17, 644a24–37. In the latter passage Aristotle associates the substantial beings of the living world with 'the ultimate species' (*ta eschata eidê*), 'the atomic species' (*to tô eidei atomon*), and 'the indivisible species' (*to eidos adiaphora*). This is also what he means by 'the particular' at (e.g.) *PA* 644a31.

[4] *Ph.* I 7, 190b25–6 (cf. 190b15–16, *Metaph.* 984a22–5, 1014b27–30). *Ph.* 191a7–12 would seem to go against this idea. There Aristotle refers to 'the underlying nature' (*hê hupokeimenê phusis*) as something that can be known only by analogy, which is supposed to be related to 'a substance possessing determinate being (*ousian echei kai to tode ti kai to on*)' as 'bronze is to the statue or wood to the bed or the shapeless prior to receiving shape to what has shape'. I confess to being at a loss about how to fit this cryptic idea into my analysis.

generation in the sense of its underlying subject (cf. *Ph.* 192a30–3). According to the *GA* I, this seed is itself a composite of matter from the female and form from the male (cf. *GC* 317a24–5). On the picture I am suggesting, what *GC* 335a32–b1 is effectively saying is that it is the material element of that seed (the species-specific blood that has its origins in the female) that is the source of its capacity to be the substance at the end of the change. Substantial generation is the fulfilment of *that* capacity (*Ph.* III 2–3).

If the above analysis is correct, then the essence of Aristotelian matter (its most fundamental feature) is not captured by its role as the subject that undergoes change.[5] For what explains the fact that material composites are capable of undergoing change, and thus what explains matter's role as the subject of change, is the fact that matter itself is a capacity to be and not be (relative to a given category of being). It is this capacity concept of matter that picks out its essence in the sense of its most causally basic feature.

4.1.2 Form

GC II 9 also extends the role of form in generation beyond what we have seen thus far. Within the context of the *Physics* I model, form is characterised simply as one of a pair of contraries, namely, the positive state that is acquired (and lost) in the process of coming to be (and passing away). In *GC* II 9 form comes to take on a teleological role as the end or goal for the sake of which generation takes place. This makes form a terminus or limit (*peras*) in the normative sense specified at *Metaphysics* 1022a4–13:

> We call a limit the last point of each thing, i.e. the first point beyond which it is not possible to find any part of the thing and within which every part is contained . . . The term is applied [in one sense] to the end (*telos*) of each thing. And this is the sort of thing *towards which* motion and action are directed, not that *from which* they proceed (though sometimes it is both that from which and that towards which, i.e. the end for the sake of which).[6]

[5] See Ebrey (2007, 2018).

[6] Aristotle may have in mind here cyclical processes that "return upon themselves" (e.g. when human generates human), where the form is both the point *towards which* one cycle advances and the starting point *from which* the next cycle proceeds. Aristotle's point is that even in these cases the form serves as a limit in the normative sense of "that for the sake of which" only in its role as that towards which the change progresses. I discuss the cyclical nature of substantial generation in Chapter 9.

It is this normative role that Aristotle has in mind in *GC* II 9 when he speaks of form as a cause of natural generation in the sense of the end for the sake of which the change takes place (335b6–7).[7]

Aristotle says in the same breath (b6–8) that form is also that feature specified in the account of each thing that makes it the kind of substance it is (*touto d' estin ho logos ho tês hekastou ousias*). But this is not the role that forms play in the context of substantial generation. While nothing crucial turns on this point, what I want to argue here is that forms play a different causal role depending on whether we are talking about *being* or *becoming* a substance. For example, the capacity for sensation is the form specified in the definition telling us what it is to be an animal, which is its so-called formal cause (*Ph.* II 3, 194b26–9). In this role the form serves as a cause of *being* an animal insofar as its presence in a living body is what makes that body (causes it *to be*) a substance of that kind.[8] But this same form also serves as a final cause of *becoming* an animal, not by its presence in the matter, but by being the goal towards which that process is directed. Here the capacity for sensation serves as a *final* cause of generation that explains why an animal embryo develops in the way that it does (e.g. why it develops eyes rather than, say, leaves).[9] If this is right, then when Aristotle says at 335b6–7 that the form is a cause and principle of natural generation 'as that for the sake of which' and that 'this is also the account of the substantial being of each thing', these are meant to pick out the form in two different roles. The latter picks out the form in its role as a formal cause of *being* a substance, while the former picks out this same form in its role as a final cause of its *becoming* one (*DA* II 4, 415b8–27). The central books of the *Metaphysics* are primarily interested in the contribution that form makes to the substantial being of sensible substances (e.g. *Metaph.* VII 17), whereas *GC* II 9 is interested exclusively in the role that form

[7] See also *Metaphysics* V 4, 1015a10–11 where nature refers to 'the form and substance' (*to eidos kai hê ousia*) which is 'the goal of its generation' (*to telos tês geneseôs*). Form is given this normative role in *Physics* II (e.g. 194a27–33), though *Physics* I 9 may already anticipate that idea (192a20–4). Kelsey (2015) argues that the central aim of *Physics* II is to establish the thesis that form is a principle of movement in the sense of a final cause.

[8] Compare *Metaphysics* VII 17, 1041b4–11 and *DA* II 4, 415b12–14.

[9] Since Aristotle's "cause" (*aitia*) is broader than our efficient cause, it is appropriate in both of these cases to speak of the relation between the form and that of which it is an *aitia* as a proper causal relation. Thus, we can say both that the form *causes* a thing to be an *F* by virtue of its presence in the matter and that it *causes* it to become an *F* by virtue of being the end for the sake of which its generation occurs. This does not commit Aristotle to saying that all formal and final causation can be reduced to efficient causation. This is especially important in the latter case: Aristotle does not invoke the notion of backwards causation; the form of the product does not "reach back" into the past to efficiently cause its own generation (Aristotle is clear about this at *GC* 324b13–18: translated below).

plays in substantial generation. And what *GC* II 9 says is that forms cause generation by being the goal or end towards which the process is directed, which explains why the substance comes to be in the way that it does (cf. *PA* 640b1–3: 'why this part comes to be first, and then that one').[10]

4.1.3 Moving Causes

By examining a parallel passage in *Metaphysics* I 3, I shall argue that the third principle Aristotle is after in *GC* II 9 (at least in the case of biological generation) is a goal-directed moving cause – the sort of principle that is 'at the same time the cause of beauty and the sort of principle from which the motion that belongs to things originates' (984b21–2).[11] It is somewhat surprising to find Aristotle arguing that matter and form alone are inadequate to explain natural generation and that we must instead have recourse to this third principle over and above those two. For, in Aristotle's own theory, the form of a thing is often pressed into the dual role as the end for the sake of which the process occurs *and* the primary moving cause that directs the process towards that end. For example, in *DA* II 4 the soul is said to function as the formal, final, *and* efficient cause of the body (415b8–27), which illustrates the point in *Physics* II 7 that these three modes of causation often converge on the same item (198a25–7).[12] And so we might expect him to treat form and matter as sufficient for generation. There should be no need to posit a separate moving cause over and above these; for one and the same ontological principle (the form) can perform three different causal roles. In order to make sense of this, *GC* II 9 must mean to treat the form and the mover as two distinct principles of generation. This is borne out by the text. First, Aristotle uses strong language to distinguish the three principles: the third principle is said to be necessary 'over and above' (*prosuparchein*, 335a31) and 'in addition to' (*proseinai*, 335b7) the matter and the form. Second, *GC* II 9 uses *eidos* interchangeably with *morphê*. And as far as I can tell Aristotle never uses the latter term when talking about the form in its role as efficient cause. Finally, in *GC* I 7 Aristotle warns against confusing forms as end states with the principle that generates the end:

[10] See also *Metaph.* 1015a10–11, *PA* 640a10–18, and *GA* 778b3–6.

[11] "Beauty" here is associated with order, arrangement, and goodness. I will have more to say about this below.

[12] Aristotle explicitly identifies the moving cause of natural generation with the organism's formal nature at *Metaphysics* VII 7, 1032a24–5.

> That which produces is a cause in the sense of whence comes the starting point of the change. But that for the sake of which is not productive of anything . . . For when the agent is present, the patient comes to be *F*. But when states are present, the patient no longer comes to be *F* but at that very moment is an *F*. And forms (i.e. ends) are a kind of state. (*GC* I 7, 324b13–18)

In what follows I will use "form" consistently to pick out the mature state of the finished product that serves as the goal of the process leading up to it (cf. *Metaph.* VII 8, 1033b5–6). On the interpretation adopted here, Aristotle treats this principle as distinct from the primary moving cause responsible for initiating the change that produces that form.[13]

4.2 Formalist Theories of Natural Generation

In order to show that matter and form are not sufficient to account for natural generation Aristotle divides previous theories into those that attempt to explain generation by appealing exclusively to forms and those that say matter alone is sufficient to generate things because the motion originates in the matter (335b16–17). The bulk of my analysis will focus on the latter. But I begin with a brief look at the former.

Aristotle provides two main arguments against purely formalist theories, both of which target the theory in the *Phaedo*. At *Phaedo* 96a1 (Plato's) Socrates turns to the causes of generation and destruction. After rejecting various theories offered by his predecessors, he eventually settles on his own Forms as the only viable causes (99d5ff.). As Aristotle reads the *Phaedo*, Socrates wants to argue that Forms cause particular things to come into being in virtue of participation (335b11–16). His objection to this sort of theory is twofold. First, if particulars come into being by participating in a Form, then everything should be in a perpetual state of generation since Forms are everlasting and there are always things participating in them. But we see that the process of generation does not go on indefinitely but has a definite stopping point. For example, there is a definite end to the generation of a human; once the offspring has acquired its adult form it ceases *coming to be* a human and at that point *is* a human. So participation in the Form Human cannot be a cause of becoming a human, since each

[13] *Meteorologica* IV also distinguishes the agent that generates a compound from the form acquired as a result of the change, which Aristotle treats as two distinct causal principles of natural generation; see 382a27–b1 (translated below). As we shall see, Aristotle often uses the expression "formal nature" (*hê kata to eidos phusis*) for both of these principles. However, even in that context he is careful to distinguish the formal nature 'as mover' from the formal nature 'as end' (*PA* I 1, 641a22–33). See Chapter 7.

particular human continues to partake of that Form long after it ceases coming into being.[14] Second, even if generation does involve participation in a Form, observation confirms that coming-to-be requires a separate moving cause: 'Yet in some cases we see that something else is the cause [of coming-to-be]; for the doctor produces health and the scientist implants knowledge, even if Health Itself and Knowledge Itself exist as well as their participants' (335b20–3).

Aristotle has often been criticised by scholars for his treatment of Plato in *GC* II 9.[15] First, they argue that Aristotle simply assumes that by identifying the Forms as the things responsible for (*aition*) generation and destruction Plato means to treat them as efficient causes, which he (Plato) clearly did not.[16] Second, Aristotle is criticised for ignoring the *Timaeus* where Plato explicitly recognises the need for a moving cause over and above the Forms.[17] Both of these criticisms misunderstand Aristotle's project in *GC* II 9. Aristotle argues that natural generation requires three principles: matter, form, and a moving cause. He targets purely formal theories of generation precisely because they do not posit this third principle in addition to forms. So clearly Aristotle does not think the *Phaedo* treated Forms as efficient causes but assigned them some other causal role in the process of generation. That is why he thinks the *Phaedo* fails as an account of generation: Socrates in the *Phaedo* omitted any mention of moving causes over and above his Forms but thought his Forms alone were adequate for the job. If Aristotle assumed that Plato meant to treat Forms as moving causes in the *Phaedo*, his objection would fall flat. The second criticism also misses Aristotle's point. While it is true that Plato posits a moving cause in the *Timaeus* over and above the Forms, *GC* II 9 is not meant as a general criticism of Plato's settled views on generation across all his works. It is a more localised attack on the *Phaedo* because the *Phaedo* is an example of the sort of theory that relies exclusively on forms as causes of natural generation. (That is why Aristotle says, 'Socrates in the *Phaedo*', and not simply, 'those who posit Forms' or 'the

[14] This is how I read 335b18–20. I take it Aristotle is referring to the generation of some particular individual here (e.g. Socrates) rather than the process of coming-to-be as such. For he thinks the latter is perpetual and everlasting. For example, while the generation of every particular animal is a rectilinear motion with a definite beginning and end, the generation of animals as such continues indefinitely without end (*GC* II 10). See Chapter 9.

[15] For modern versions of this criticism see Vlastos (1969) and Annas (1982).

[16] See Philoponus (*in GC*, 281,25–6; 285,13–4), who takes it for granted that Aristotle is thinking of Forms as efficient causes here. Modern commentators follow him.

[17] I discuss the *Timaeus* model in Chapter 8.

friends of the Forms'.) The fact that the *Timaeus* is not vulnerable to Aristotle's criticisms is irrelevant to the dialectical structure of *GC* II 9.

4.3 Materialist Theories of Natural Generation

After criticising the *Phaedo* approach to generation, Aristotle turns to the materialists who claim that matter alone is sufficient to explain the change because the process originates in the matter (*GC* 335b16–17).[18] Aristotle is certainly sympathetic to the materialists' project here. For it is 'more naturalistic' (*phusikôteron*) to attribute generation to the motions of matter than to separately existing Forms since what alters and transforms a thing is 'more [sc. efficiently] causally responsible' (*aitiôteron*) for its generation (335b24–8). Nevertheless, he thinks that purely materialist accounts ultimately fail as well.

As a way into Aristotle's arguments against materialism in *GC* II 9 we can note how closely they resemble those in *Metaphysics* I 3. The stretch of text that I want to focus on here runs from 984a16 to 984b22. It is here that Aristotle begins to make his case for the need to posit a special kind of efficient cause just like *GC* II 9. And like *GC* II 9 (see 335b29–33) he starts off by targeting those who only recognised the material subject as a cause:

> From these facts one might think that the only cause is the so-called material cause. But as men thus advanced, the very facts opened the way for them and joined in forcing them to investigate the subject. However true it may be that all generation and destruction proceed from some one thing, or from more than one, then why does this happen and what is the cause? For at least the subject (*to hupokeimenon*) itself does not make itself change, e.g. neither the wood nor the bronze causes the change of either of them, nor does the wood manufacture a bed and the bronze a statue, but something else is the cause of the change. And to seek this is to seek the second cause, as we should say, whence comes the origin of motion (*hothen hê archê tês kinêseôs*). (984a16–27, Ross translation with modifications)

After briefly discussing how the earliest attempts to grasp the efficient cause were all confused, Aristotle goes on to say that those who made the elements (or simple bodies)[19] more than one and endowed some of them

[18] By saying generation 'derives from' (*apo*) the matter I take it Aristotle means to say that the materialists treated matter as the primary efficient cause of the change, i.e. 'that which contains the origin (*archê*) of the change' and is 'first among the causes' (*GC* I 7, 324a26–8). Aristotle's objections to this thesis do not appear to rely on his technical notion of matter as a capacity to be and not be, though as we shall see he does invoke matter as subject (at least in the first objection).

[19] I will use 'the elements' and 'the simple bodies' interchangeably in this book to pick out the most basic material components of all physical bodies. Aristotle often includes here earth, air, fire, and

with a 'motive nature' were in a much better position to give an account of it:

> But for those who made things more than one (e.g. hot and cold or fire and earth) it was possible to state it; for they made use of fire as having a motive nature (*kinêtikên echonti tô puri tên phusin*), while to water and earth and other things of this sort they gave the opposite nature [sc. the nature of a subject]. (984b5–8)

This passage makes it clear that Aristotle does not think all materialists failed to recognise the need for an efficient cause. While some of them clearly omitted this principle (984a16–27), he is explicit here that the more sophisticated brands of materialism give some of the elements the capacity to effect change. (This is important since, as we shall see, it is this more sophisticated materialist that Aristotle targets at the end of *GC* II 9.) Finally, even this version of materialism ultimately fails because the motions of the simple bodies alone are not adequate to account for the beauty, order, and goodness exhibited by natural generation:

> After these thinkers and their inquiry into these sorts of principles, because they were inadequate for generating the substance of existing things, men were once again compelled, so to speak, by the truth itself to inquire into the next principle. For it is not likely that fire or earth or any other such being could cause things to be or come to be in a good and beautiful condition (*eu kai kalôs*), nor indeed that those thinkers should suppose such a view. Nor again was it right to entrust the fact that things are in a beautiful state (*pragma kalôs eichen*) to spontaneity and chance. So when someone said there is intelligence in nature just as in animals and that this is the thing responsible (*ton aition*) for the order (*kosmou*) and arrangement (*taxeôs*) of everything, he seemed like a sane man in contrast to the random statements of his predecessors. We know that Anaxagoras certainly adopted these views, but Hermotimus of Clazomenae is credited with expressing them earlier. Those, then, who conceived of things in this way stated that there is a principle of things that is at the same time the cause of beauty and the sort of cause from which the change that belongs to things originates. (984b8–22)

water, which he calls 'the simple bodies' in the *Physics* (e.g. 192b10–11, 214b13) and 'the so-called elements' at *PA* 646a13. Some Presocratics also included things like the hot and the cold, and we might also include *pneuma* (e.g. *GA* 741b25–742a16, 789b10–12). When talking about 'elemental transformation' we are talking about the mutual transformation of these so-called elements (earth, air, fire, and water) into one other. In Aristotle's own system, these simple bodies are themselves constituted by lower-level forces that he calls the 'elemental powers' (*dunameis*), namely, hot, cold, moist, and dry, which he treats as the most basic constituents of all physical bodies (*PA* 646a14–16).

The last sentence of this passage makes it clear that we are still talking about the efficient cause of generation (rather than, say, the final cause). The upshot of the passage is that a proper scientific account of the natural world requires not just positing a source of motion (the materialists at 984b5–8 clearly did that) but one whose very nature is to produce beauty, order, and goodness. Otherwise we cannot account for this aspect of the natural world except by attributing it to chance, and Aristotle thinks these features cannot be the product of chance. As we shall see in Chapter 8, Aristotle shares this commitment with Plato. Aristotle, like Plato in the *Phaedo*, praises Anaxagoras because he thinks that intelligence is the sort of cause that aims at the good by nature and so can serve as a principle of the sort he is after. But unlike Plato Aristotle thinks that there are other order-producing causes besides reason.[20] Before returning to *GC* II 9, we need to say something about Aristotle's use of *to kalon* in this context.

In some contexts *to kalon* means beautiful in an aesthetic sense, where it refers to things whose features are pleasing and attractive to the senses (e.g. a beautiful painting or sculpture).[21] In *PA* I 5 Aristotle says that his use of *to kalon* in relation to the works of nature (specifically living nature) is instead connected with their teleological structure:

> So too one should approach research about each of the animals without disgust, since in all of them there is something natural and beautiful (*kalon*). For what is not haphazard but occurs for the sake of something is in fact present most of all in the works of nature. And in this domain the end for the sake of which each animal has been constituted or comes to be takes the place of the beautiful. (645a24–6)

In this passage Aristotle argues that the *kalon* is present also in the works of nature. It is clear from the context that he does not mean beautiful in the aesthetic sense; rather, he means 'what is not haphazard but occurs for the sake of something'.[22] Aristotle is responding to those who might dismiss the study of living things as something ignoble and physically repulsive (who 'recoil with childish disgust'). His response is that biology is not disgusting, that there is beauty even in the messy guts, internal organs, and the lowly creatures ('in all of them there is something natural and beautiful'), and that in this context the beauty is to be found in the way

[20] The phrase "order-producing cause" comes from Johansen (2004, e.g. 72). I will have a good deal more to say about Plato's influence on Aristotle's model of generation in Chapter 8.

[21] For a discussion of this sense of *to kalon* see Irwin (2010, 384–5).

[22] Compare *Metaph.* 1078a31–b6. There Aristotle identifies 'order (*taxis*), symmetry (*summetria*), and definiteness (*hôrismenon*)' as forms or species (*eidê*) of *to kalon*.

animals and their parts are specially adapted for the sake of an end (their structure is fitting or appropriate for their functions). Here to say that things come to be 'in a good and beautiful condition' is to draw attention to the fact that living things and their parts exhibit the sort of adaptive complexity that promotes their survival and well-being.[23]

While the *PA* restricts this analysis to living nature, it is clear that *Metaphysics* 984b8–22 also means to include under the beautiful and the good the orderly motions of the heavenly bodies and the regular cycle of the seasons that follow upon them, what Plato includes among 'the greatest and most beautiful things' (*Laws* X, 889a4; cf. *PA* 641b11–23). Scholars remain divided over whether Aristotle accepts that the same teleological adjustment of means to ends characteristic of living things is also a feature of these other phenomena.[24] My sympathies lie with the view that Aristotle restricts this robust kind of teleology to organisms: only in the case of living things is it true to say that their parts exist and come to be *for the sake of* the good condition of the organism that they promote.[25] On this reading, the mutual transformations of the elements and the various meteorological phenomena they underwrite (e.g. the cycle of evaporation and condensation that produces annual rainfall) display beauty and goodness only in the weak sense of exhibiting a kind of regular order rather than being random and disorderly. The efficient cause that is responsible for this order and arrangement is the annual rotation of the sun (*GC* 336a21–b15, *Meteor.* 346b20–347a6). But this is not the kind of order-producing efficient cause that operates for the sake of an end and so does not count as the sort of efficient cause that *GC* II 9 is after.[26] In what follows I will focus exclusively on the generation of living things and their parts, since it is uncontroversial that in these cases Aristotle thinks we must have recourse to a goal-directed efficient cause to explain the change. With these texts as background we can now return to *GC* II 9.

GC II 9 raises three main objections to those purely materialist accounts that treat matter as sufficient for generating things. First Aristotle argues that matter cannot be solely responsible for natural generation because matter is the patient that undergoes the change:

23 For a similar reading of *to kalon* in biology see Irwin (2010, 386–88). I disagree with Irwin (388) that beauty is an inappropriate translation for the biological use of *to kalon* as long as we allow that the English word "beauty" is flexible enough to accommodate a non-aesthetic sense, which it surely is. For example, Sylvia Nasar's 'A Beautiful Mind' is not a reference to the physical attractiveness of J. F. Nash Jr's brain (its pleasing colour or shape) but to his intellectual acumen.

24 Cf. *GC* 337a1–7. See Scharle (2015, e.g. 88–9).

25 I argue against the Cosmic Teleology view in Henry (2015b), which is the ancestor of Chapter 9.

26 For more on this see Chapter 9.

> But what they say is not right. For matter is what is acted upon and moved (*to paschein esti kai to kineisthai*), while effecting motion and acting belongs to another power (*to de kinein kai poiein heteras dunameôs*). This is clear both from cases of craftsmanship and from natural generation. For water itself does not produce an animal from itself, nor does the wood make itself into a bed, but rather the art does this. (335b29–33)

This is essentially the same argument that Aristotle deploys at *Metaphysics* 984a16–27. In both places Aristotle is focused on matter as *the passive subject* of change that is acted upon and moved. But generation also requires an efficient cause to set the process in motion, which he says here belongs to 'another power'.[27]

Aristotle's second objection (which does not have an analogue in *Metaphysics* I 3) holds that by eliminating the form the materialists omit 'the more controlling cause' (*tên kuriôteran aitian*) (335b33–5). On Joachim's reading Aristotle is complaining about the fact that the materialists omitted reference to an efficient cause. But this is not quite right. What the materialists are said to omit here is 'the essential form' (*to ti ên einai kai tên morphên*). And Aristotle has already said that in the context of generation forms are causes as ends not as movers (335b6–7). This suggests that Aristotle is targeting that fact that the materialists failed to avail themselves of the *final* (not efficient) cause of generation.[28] But in what sense is this cause 'more controlling' (*kuriôteran*) in generation? In *PA* I 1 Aristotle argues that final causes are primary to efficient causes in coming-to-be on the grounds that the account of the product's form, which specifies the end to be achieved, is an *origin* or *starting point* of change:

> Since we see more than one cause of natural generation, e.g. both that for the sake of which and whence comes the origin of motion, we also need to distinguish concerning these which is naturally prior and which is posterior. Now it is clear that the one we call that for the sake of which is prior; for this is an account, and the account is a starting point (*archê*) among things put together by craftsmanship and nature alike. For once the doctor has defined health and the builder has defined a house, either by thought or perception, they provide the accounts and the causes of each of the things they produce and give the reason why it must be produced in this way. And yet that for the sake of which and the good are present more in the works of nature than those of craftsmanship. (*PA* I 1, 639b11–21; cf. 646a24–b9)

27 I return to this objection at the end of the chapter. The issue is whether Aristotle is committed throughout his works to the view that matter as such is something passive and inert (as this objection suggests) or whether he came to think of matter as a source of active motions in its own right.

28 This reading is also adopted by Philoponus (*in GC*, 287,12–15).

The idea here is that the form of the product causes generation, not by initiating the change (efficient causation), but by being the goal towards which that change is directed (final causation). It is in this sense that the form (as goal) serves as a starting point of generation. For example, when constructing a house, the builder starts by considering the form of the house she wants to build and then coordinates her actions in a way that tracks that goal. This makes the form the more controlling cause in generation: while the agent is responsible for physically transforming the matter, it is the form (serving as a target) that ultimately explains why that agent behaves in the way that it does (why it produces *this* before *that*).[29] By eliminating any reference to the form as a goal of the process, the materialists are thus once again left without the resources necessary to explain generation.

Aristotle had already made a similar point against Empedocles back in *GC* II 6 (333b3–24; cf. *Ph.* II 8, 198b16–199a8). According to Empedocles' theory, the parts of living things are constituted by a specific ratio of elements. Bone, for example, comes to be when the elements are put together according to the formula $E_2W_2A_0F_4$. That formula specifies the form that makes those elements bone (333b13–15). Since the formation of bone is a regular feature of a species' development, there must be some reliable cause at work in the process that puts the elements together in those precise ratios. Chance cannot be responsible for that since chance is too irregular. This is where the trouble starts for Empedocles:

> For it [bone] does not come to be from the elements being put together in any chance way but, according to his own statement, only when they are put together in a certain ratio (*logos*). What, then, is responsible for this [sc. putting the elements together in that specific way]? For it is not fire or earth. But neither is it Love or Strife, for Love is a cause of coming together only while Strife is only a cause of coming apart. (333b9–13)

According to this argument, Love and Strife cannot be primary causes of bone development since they are causes of combination and separation only, not combining and separating elements in any specific way. This is what Aristotle means at 333b22–4 when he says that the motions of Love and Strife are too simple to be primary causes of natural generation: 'For it is not enough to say that Love and Strife set things in motion unless it is the nature of Love to be a cause of *this* kind of motion and Strife *that* kind of

[29] Compare *GA* I 22, 730b8–22. Aristotle does not say how the form of a natural process can serve as a goal without being the object of a desire. But that is not relevant for my purposes here. For a discussion of this point see Henry (2017).

motion.' When it comes to the formation of bone what needs to be explained is not simply the fact that the elements are combined but that they are regularly combined in the precise ratios $E_2W_2A_0F_4$. Since Empedocles treats Love as a source of coming together only not coming together in any specific way, it is fair to say that while Love is an intrinsic (*per se*) cause of the fact that the elements are combined it is only an incidental (*per accidens*) cause of their being combined in that specific way – and so only an incidental cause of there being a bone at the end of the change. This makes the formation of bone a chance event (*Ph.* II 5–6). And the process is far too regular for that.

To put the point slightly differently, Empedocles' efficient causes do not aim at combining and separating the elements according to any precise ratios; they are not, as Aristotle puts it at *GC* 336a1–2, the sorts of causes whose nature is to initiate motion 'according to the form' (*tên kata to eidos aitian*) of the product. As a consequence, Empedocles has no choice but to ascribe organogenesis to chance. Love is like the blind cook who throws ingredients together into a pot: It is possible that blindly throwing ingredients together will result in a gourmet dish by chance. But we would certainly not expect to find this method yielding that result either always or for the most part. The kind of regularity and order that is characteristic of expert cooking depends on combining ingredients according to the proportions specified by a recipe. Likewise, in order to account for the fact that living things regularly breed true to type (333b7–11), it is not enough for Love *merely* to combine the elements; those elements must be combined in the precise ratios that make them the sort of parts they are.[30]

Aristotle goes on to say that the cause in question is the substantial being (*ousia*) of each thing, which he identifies with the product's nature and good (333b13–18). In doing so he is most likely pointing to the role that forms play as *final* causes of generation (cf. 335b6–7). If this is right, then Aristotle's objection to Empedocles is similar to the second objection raised against the materialists in *GC* II 9. By failing to make use of the ratios of things (their essential form) as final causes of the motions initiated by his efficient causes (Love and Strife), Empedocles ends up ignoring the more controlling cause of natural generation. As a result, the fact that living things and their parts are constructed in way that promotes their good turns out to be an incidental side effect of the actions of his moving causes and not the end for the sake of which they operate.

[30] Cf. Sorabji (1980, 179).

Aristotle's third objection to the materialist thesis is difficult to unpack, and so it is worth quoting the passage in full:

> Further, since they remove the cause that operates according to the form (*aphairountes tên kata to eidos aitian*), they give the simple bodies generative powers that are too instrumental (*tas dunameis apodidoasi tois sômasi, di' has gennôsi, lian organikôs*). For since they say that the hot naturally divides things and the cold naturally amalgamates them, and that of the rest of the contraries some act while others suffer action, they maintain that it is out of such things and by their agency that everything else comes to be and passes away. Yet it is obvious that even fire itself is moved and suffers action. And, further, they act in virtually the same way as one who treats the saw and the other tools of carpentry as the cause of the things that come to be; for the wood *must* be divided if there is sawing and *must* become smooth if there is planing, and so on with the rest of the tools. Hence, even if it is true that fire most of all acts and effects motion, there is a further point which they fail to observe, namely, that it is worse than the <craftsman's> tools in how it sets things in motion. (*GC* II 9, 336a1–12, translated after Joachim)

All three of the objections against materialism in *GC* II 9 target the thesis (mentioned first at *GC* 335b16–17) that matter alone is sufficient to generate things because the motion originates in the matter. I suggested above that Aristotle's first objection (at *GC* 335b29–33) is aimed at those materialists mentioned at *Metaphysics* 984a16–27 who treated matter exclusively as a subject of change. Here he turns to that other kind of materialist mentioned at *Metaphysics* 984b5–8 who assigned things like fire or the hot and cold 'a motive nature' in virtue of which they effect change. According to this more dynamic brand of materialism, among the elements that constitute the matter of natural bodies, some are active while others are passive, so that 'it is out of such things *and* by their agency that everything else comes to be and passes away' (336a3–6). This allows the materialists to explain natural generation in terms of matter alone while avoiding Aristotle's first objection.[31]

By saying that these materialists removed 'the cause the operates according to the form (*tên kata to eidos aitian*)' Aristotle means that they eliminated from their account the kind of moving causes that direct the

[31] Joachim (1926, 251). Elsewhere Aristotle refers to the elements or simple bodies as 'the matter of composite bodies' (*PA* 646a13–17; see also *De juv.* 466a21–3). And so we can treat their motions as 'motions of matter' here, even though Aristotle has just said that matter qua matter is passive (*GC* 335b29–33). It is a further question, to which I return below, whether the elements count as 'matter' when operating as efficient causes.

process towards the form.[32] Instead they ascribed natural generation to the simple bodies and their inherent powers (e.g. fire, the hot, and the cold). Aristotle objects that the generative powers they give to these agents are 'too instrumental' (*lian organikas*) to do the job required of them.[33] He makes a similar point about Democritus' account of teeth-shedding at *GA* V 8. There he argues that, even if the sorts of necessary causes that Democritus invokes are at work in the process of shedding teeth, they play the same role as the craftsman's tools (789b3–15; cf. 740b25–741a3, 743a36–b18), and so to try and explain the process in terms of those causes alone is like trying to explain why the patient was healed by reference to the lancet alone while ignoring the surgeon who made the incision for the sake of health. In this sense, Aristotle says in our *GC* passage, the materialists 'act in virtually the same way as one who treats the saw and the other tools of carpentry as the cause of the things that come to be' (336a7–9; cf. *Timaeus* 46c10–12).[34]

While it is not clear how to fit together Aristotle's various remarks in *GC* 336a1–12, they all appear to be directed towards a single overall point. By treating natural generation as the effect of various elemental forces, the more dynamic brands of materialism end up treating as *primary* causes (324a26–8: 'that which contains the origin of the change' and is 'first among the causes') what are at best only *proximate* causes (324a32: 'the last mover' that 'imparts motion by being itself moved'). As we have seen, Aristotle thinks that an adequate theory of generation must be able to explain the fact that naturally generated substances exhibit order, beauty, and goodness, all of which he thinks are beyond the simple natures of the elements themselves (*Metaph.* I 3). As he puts it, their generative powers are 'too instrumental' to be primary causes of those features. Like the tools of craftsmanship, such causes can only produce those effects consistently when operating under the guidance of some goal-directed efficient cause that aims at the good by nature.[35] This seems to be what Aristotle has in

[32] Aristotle uses similar terminology for the efficient cause of natural generation at *Metaph.* VII 7, 1032a24–5 (*hê kata to eidos phusis*), which he lists alongside matter and form.

[33] This reading was suggested to me by Emily Katz (2017 and personal communication). Compare Joachim (1926, 252), Williams (1982, 185), and Kelsey (2003, 83). A similar reading was suggested by Emily Kress in her commentary on a draft of this chapter that I presented to the ancient philosophy reading group at Yale (April 2017) as well as by James Allen (personal communication).

[34] As we shall see, by calling things like heat and cold "instruments" or "tools" (*organa*), Aristotle means to downgrade their role in generation from that of *primary* causes to what he calls *co*-causes (e.g. *DA* 416a13–18). This forms the basis of what in Chapter 8 I call the architectonic model of generation, which I argue has its roots in Plato's *Timaeus*.

[35] See *GA* 734b31–735a4, or the more generous *Meteor.* 390b2–14.

mind in the closing lines of our passage when he says that fire is inferior to the craftsman's tools in how it sets things in motion. Fire's motion on its own is too simple and unstructured compared to the tools of the craftsman, which owe their own organised movements to the activity of the craft (*technê*).[36]

Thus, whereas the first group of materialists failed to posit any efficient causes at all but instead conceptualised matter simply as the subject of change, these more sophisticated materialists recognised the need for another principle to act as a source of motion. But their efficient causes (the simple bodies and their simple powers) are 'too instrumental' to account for natural generation. According to *Metaphysics* I 3, we must have recourse to some other principle that is 'at the same time the cause of beauty and the sort of cause from which the change that belongs to things originates'. This makes sense of Aristotle's more general complaint in *GC* II 9 that this third principle was something 'vaguely dreamed of by all' but 'definitely stated by none' (335b7–8). Aristotle is not saying that the materialists all failed to posit efficient causes of natural generation (some of them clearly did) but rather that what they offered was an impoverished version of one. This is why Aristotle praises Anaxagoras in *Metaphysics* I 3 since intelligence (*nous*) is the sort of cause that can generate those features of the world that he thinks an adequate theory of natural generation must explain (its order, beauty, and goodness). Aristotle goes on to reject Anaxagoras' account because in the end he too failed to invoke *nous* as a cause of generation but instead fell back on more mechanical causes characteristic of the other materialists.

Here I agree with Gotthelf's (2012, 24) assessment of the motivation behind Aristotle's natural teleology: 'Given the simplicity of Aristotle's chemistry, he can only believe that the outcome of organic development is too complex, too orderly, possessing too much of limit, *logos*, and form, to be the result merely of the unlimited, relatively, indefinite natural action and interaction of the elements.' Even if the more dynamic brand of materialism recognised the need for efficient causes, what is missing from their account is the sort of cause whose nature is to produce beauty, order,

[36] Aristotle makes this very point at *DA* II 4, 416a13–18: 'While fire is, in a way, a co-cause (*sunaitian*) [sc. of biological growth], it is not the cause in the unqualified sense, which is more properly the soul. For the growth of fire goes on without limit as long as there is fuel to burn, while in everything constituted by nature there is limit and proportion to their size and growth. And this is due to soul, not fire, and to the form more than the matter.' We shall have a chance to return to this passage in Chapter 8.

and goodness. As we shall see in Chapter 8, this is precisely the role that Aristotle assigns to the formal natures of living things.

To close this chapter, I want to return to Aristotle's first objection against materialism at *GC* 335b29–33 since it raises more questions about his views regarding the role of matter in natural generation and whether those views are consistent with what Aristotle says about generation elsewhere (especially in the biological works).

4.4 Is Matter Entirely Passive and Inert?

The materialist thesis under consideration in *GC* II 9 holds that matter is sufficient to account for generation because the process originates in the matter (335b16–17). As we have seen, Aristotle objects to this thesis in part because he thinks matter is the patient in generation and as such is acted upon and moved, whereas acting and effecting motion belong to another power (b29–33). By contrast, the more dynamic version of materialism that Aristotle targets at 336a1–12 treats natural generation as an effect of those elemental forces that constitute the matter of natural bodies. Joachim characterises this version of materialism (which he attributes to the Pythagoreans) as follows:

> The Pythagorean materialists regard *genesis* and *phthora* as the effects of certain forces – e.g. the hot and the cold – inherent in, and constituent of, the matter of which bodies consist. It is the nature of each of these "elementary qualities" or "material forces" to act or suffer action in certain definite ways. Hence the hot and the cold, and the like, are *both* the materials out of which (or into which), *and* the forces by means of which, all the other things come-to-be (or pass-away). (Joachim 1926, 251)

Since Aristotle rejects this form of materialism, it is reasonable to assume that he is thinking of matter in *GC* II 9 not as a dynamic force capable of effecting change on its own, but as something entirely passive and inert, like the carpenter's hammer lying motionless on the table waiting to be moved by his craft.[37] The problem is that this way of conceptualising matter appears to conflict with Aristotle views outside *GC*. For example, in *Meteorologica* IV Aristotle develops a theory of generation that looks a lot like the dynamic version of materialism targeted in *GC* (and *Metaph.* I 3). For both characterise natural generation as the effect of certain lower-level forces inherent in, and constituent of, the matter of which bodies consist

[37] Cf. *Metaph.* XII 6, 1071b30–1: 'Matter will surely not move itself, rather the carpenter's art does this; nor do the menstrual fluids or the earth, but rather the seeds and semen move them.'

(as Joachim puts it). To appreciate the problem, we need to take a brief look at *Meteor.* IV, where Aristotle also claims to be investigating coming-to-be in the unqualified sense.[38]

In *PA* II 1 Aristotle develops a hierarchical picture of animal bodies involving three levels of composition. At the lowest level is the composition of the uniform parts (e.g. flesh, bone, sinew) out of the elemental powers (hot, cold, moist, dry) that are said to constitute 'the matter of composite bodies' (*hulê tôn sunthetôn sômatôn estin*).[39] *Meteor.* IV is focused on generation at this level (the generation of uniform natural bodies out of the elemental powers). At this level coming-to-be in the unqualified sense is said to result when the two active forces (the hot and the cold) 'prevail over' the two passive forces (the dry and the moist):

> We have explained that the causes of the elements are four and that their couplings (*suzugias*) determine the number of the elements to be four. Two of the causes (the hot and the cold) are active and two of them (the dry and the moist) are passive. (*Meteor.* 378b9–13)

> First of all, generally speaking, unqualified generation and natural change are the work of these powers (*hê haplê genesis kai hê phusikê metabolê toutôn tôn dunameôn estin ergon*), as is the opposite process of natural destruction. And these are found among plants and animals and their parts. Natural generation in the unqualified sense is a transformation brought about by the agency of these powers out of the underlying matter (*ek tês hupokeimenês hulês*) of each natural body (this corresponds to the passive powers we mentioned), when present in a certain ratio. When the hot and the cold prevail over the matter they generate a thing; when they fail to overpower it, the body is imperfectly boiled or otherwise inconcocted. (*Meteor.* 378b28–379a3)

In *Metaphysics* VII 7 Aristotle lays it down that everything that comes into being comes to be by the agency of something (efficient cause), out of something (matter), and becomes something (form) (1032a12–14; see also VII 8, 1033a24–8). This axiom supplies the organising principle for *Meteorologica* IV:[40]

[38] For a thorough discussion of *Meteor.* IV in this dynamic context see Lennox (2004b). See also the introductory essay in Lewis (2014) and Connell (2016, 72–4).

[39] *PA* 646a13–17. See also *De juv.* 466a21–3: 'These (hot, cold, moist, dry) are the matter of bodies.'

[40] This axiom also provides the starting point for a number of arguments in the *Generation of Animals* (e.g. *GA* II 1, 733b24–6). It is sometimes fleshed out by saying that what is generated is generated *by* something that already bears the form in a state of completion *out of* something that already is that thing in potentiality (e.g. *GA* II 1, 734a29–31, b20–2, b34–6). The three principles identified here correspond to the principles of generation marked out by what I am calling the "extended" hylomorphic model.

> There are two causes apart from the matter, viz. the agent and the affection (*to te poioun kai to pathos*). The agent is a cause in the sense of whence comes the change (*hôs hothen hê kinêsis*), while the affection is a cause in the sense of form (*hôs eidos*) ... As we explained, the agent operates by means of two powers and the patient is acted upon in virtue of two powers: action occurs by means of heat or cold, while the affection results either from the presence or absence of heat or cold. (*Meteor.* 382a27–b1)

According to this passage, in order to explain generation at the level of uniform natural bodies we must make reference to three causal factors: (i) the matter out of which the process occurs; (ii) the agent responsible for initiating that change; and (iii) the forms acquired as a result of the change.[41]

For example, in *Meteor.* IV 5 Aristotle says that every natural body that contains a portion of fluid (matter) becomes dry (form) when subject to heating and cooling (agents). He goes on to describe in *Meteor.* IV 6 how bodies that are either water or compounds of water and earth (matter) become solidified (form) when exposed to heating and cooling (agent). Those that are solidified by dry heat are soluble in water, while those that are solidified by cold (i.e. frozen) are soluble by heat. Of bodies that are solidified by cold and are composed of earth and water where earth preponderates (matter), those that are solidified by the loss of heat (agent) are meltable by heat (form), while those that are solidified by refrigeration where all the moisture has been vaporised, e.g. iron and horn, are insoluble except by extreme heat, though they can be softened to the point of becoming liquid and then solidifying again. In each of these cases a natural uniform body is generated by a transformation brought about through the agency of the two active powers (hot and cold) operating on the corresponding passive powers (dry and moist), which produces certain formal properties (solidity, insolubility, etc.).[42]

This should be enough to make it clear that the *Meteor.* IV account of generation shares some commitments with the more dynamic version of

[41] On the different kinds of forms that differentiate uniform natural bodies see *Meteor.* IV 8. These include both the sensible features of a body and its primary material dispositions (e.g. solidity, fusibility, solubility, etc.).

[42] Lennox (2004b) characterises the project of *Meteor.* IV 5–7 as an attempt to formulate various 'thermodynamic empirical laws'. This is an interesting idea, and I agree that the explanations there can be reformulated in terms of such laws. However, I have my doubts that this was Aristotle's intended aim. Instead, I have argued that *Meteor.* IV 5–7 is best read as an attempt to identify, at the level of uniform natural bodies, the three causal principles specified by the axiom set out at the start of *Metaphysics* VII 7–9 (matter, form, and moving cause). This axiom, I have suggested, provides the organising principle for those chapters (as stated at *Meteor.* 382a27–b1).

materialism targeted at the end of *GC* II 9. Aristotle identifies the four elemental powers with the matter of composite bodies (*PA* 646a13–17, *De juv.* 466a21–3). And he is explicit in *Meteor.* IV that one pair (the hot and cold) is active and capable of effecting change while the other pair (the moist and dry) is passive and gets acted upon (378b13–26). In this sense Aristotle agrees with those materialists who claim that, among the elemental forces that constitute the matter of natural bodies, some act while others suffer action so that it is out of such things *and* by their agency that other things come to be (at least at the level of uniform natural bodies). This is not what we should expect if Aristotle thought matter's contribution to generation was exhausted by its role as a passive subject of change, as the objection at *GC* 335b29–33 first suggests.

The *Meteor.* IV account is not the only source of tension with *GC* II 9. There are also Aristotle's remarks in the *Generation of Animals* about parthenogenesis (female induced embryogenesis). For the most part Aristotle treats semen as the agent that imparts form, while the matter from the female is the patient that is acted on and receives that form. As Aristotle puts it in *GA* I 21: the female qua female is the patient (*pathêtikon*) while the male qua male is the agent and efficient cause of the change (*poiêtikon kai hothen hê archê tês kinêseôs*), so that one thing comes to be from them in the way in which a bed comes from the wood and the carpenter or a ball from the wax and the form (729b12–18). For this reason, Aristotle says, females cannot generate on their own: 'for the matter requires a principle to trigger its motion and determine its form' (730a26–30). So far this looks like a version of the principle behind Aristotle's objection to materialism at *GC* 335b29–33. However, on several occasions Aristotle qualifies this by noting that in some cases menstrual blood is able to form itself into an embryo and catalyse its own development in the absence of any contribution from the male (e.g. 730a30–2).[43] In these cases menstrual blood is both *that out of which* and *that by means of which* the new animal comes to be. Parthenogenesis thus looks like another counterexample to the objection in *GC* II 9 that matter cannot generate on its own since it is entirely passive and needs something else to set it in motion.

Finally, there are Aristotle's explicit remarks in *Metaphysics* VII 9 that matter has the capacity to set itself in motion. He begins the chapter by asking: 'Why is it that some things can come to be both through

[43] For the philosophical significance of parthenogenesis in relation to Aristotle's reproductive hylomorphism see Chapter 6.

craftsmanship and of their own accord, e.g. health, while others cannot, e.g. houses?' The reason, Aristotle says, is that in the former case 'the matter that forms the starting point of coming-to-be' (*hê hulê hê archousa tês geneseôs*) is such that it can 'be moved by its own agency' (*kineisthai huph' hautês*) in the same way the craft moves it, whereas those materials that cannot do this need the craft to impose form on them (1034a14–19). Aristotle extends this idea to natural generation towards the end of the chapter:

> It is the same with things that are formed by nature. For the seed produces things just as those that come to be from the craft (for it has the form in capacity, and that from which the seed comes in a way shares the same name as the product) ... And, just like the formation of craft products, those things are spontaneously generated whose matter is capable of moving itself with the same motion that the seed initiates, while those whose matter is not like this cannot be generated in any other way than from others like themselves. (1034a33–b7)

According to this account, the matter from which spontaneous generation proceeds is capable of effecting the same kinds of changes that seed initiates in sexual generation.[44] This conflicts with Aristotle's statement in *GC* II 9 that matter cannot generate on its own because it is passive and inert. The conflict becomes especially stark when we consider that in *GC* II 9 Aristotle says this 'is clear both from cases of craftsmanship and from natural generation'. The whole point of *Metaphysics* VII 9 is that there are certain cases of craftsmanship and natural generation where the matter *can* bring about the product by itself. So once again it seems that Aristotle has been hoisted with his own petard.

There are two ways of resolving this apparent conflict that I do not think show much promise. One is simply to argue that Aristotle eventually changed his mind on the role of matter in generation. In *GC* he treated matter as passive and held that generation requires something else to effect motion. Later he came to embrace a more dynamic conception of matter, perhaps forced by empirical observations like parthenogenesis and spontaneous generation.[45] In the absence of any explicit statement from Aristotle

[44] It is difficult to tell what Aristotle means by saying that the matter of a spontaneous generation 'is capable of moving itself with the same motion that the seed initiates'. In the *GA* III 11 account of spontaneous generation, the sun (the analogue of the father) serves as the primary efficient cause of the change. But Aristotle also mentions a kind of heat that gets enclosed within the "embryo", which is some kind of internal efficient cause. In the end, Aristotle's remarks in *GA* III 11 are too cryptic to draw any definite conclusions about the *Metaphysics* VII 9 doctrine (though various competing interpretations abound).

[45] Connell (2001, e.g. 313–14, 321).

that he rejected or revised his concept of matter, I think we should adopt this kind of strong developmentalism only as a last resort. A second alternative is to deny that Aristotle actually endorses the concept of matter that underwrites the objection at *GC* 335b29–33. Instead he is arguing dialectically against those of his rivals who treated matter as passive and inert, without actually committing himself to that view. While I think there is some truth to this (see below), Aristotle clearly endorses that concept of matter in *GC*: at 324b18 he says, clearly in his own voice, that matter qua matter is passive (*hê d' hulê ê hulê pathêtikon*).

To get a sense of what Aristotle may have in mind at *GC* 335b29–33, I propose we start with what he says in the twin passage at *Metaphysics* 984a16–27. There we are told that the first wave of materialists were focused on matter as the underlying subject (*to hupokeimenon*) from which a thing first comes into being and into which it is eventually destroyed (983b6–9). This is inadequate to account for generation, he says, because 'the underlying subject itself does not make itself change'. What Aristotle seems to be saying here is that *being a subject* (in the *Physics* I 7 sense) implies being a patient *and only a patient* of change. This is why Aristotle thinks the second wave of materialism was an advancement in the search for causes. Those materialists recognised the need to appeal to another kind of capacity besides the passive capacity associated with being a subject of change and so endowed some of the elements with generative powers of their own (984b5–8). While Aristotle definitely thinks of matter as playing the role of *hupokeimenon* in generation, the question is whether he thinks its role is exhausted by being a passive subject of change or whether being a *hupokeimenon* is only one of its roles. This yields two different readings of *GC* 335b29–33.

On the first reading (call it Reading A) *GC* 335b29–33 is making the analytic claim that matter *by definition* cannot account for generation since the essence of matter is exhausted by its role as the patient of change.[46] On this reading, matter is characterised entirely by its passive capacity to be acted upon and moved. In that case we clearly cannot explain generation in terms of matter alone since generation requires both a patient and an agent. (This is how I presented the objection above.) Since matter as such is acted upon and moved, we have to appeal to something besides the matter that has the capacity to effect motion. On the alternative reading (Reading B) matter is a patient only in its role as subject, where the latter concept

46 Ebrey (2018).

(rather than the concept of matter itself) implies being the thing that gets acted upon and moved. This leaves room for a more dynamic conception of matter that endows it with other powers besides the passive capacities it has in virtue of being a subject.[47] Read in this way all Aristotle needs to be saying at *GC* 335b29–33 is that any version of materialism that only conceptualises matter as a subject of change (as some of his predecessors clearly did) will not be able to explain natural generation, since generation requires the power to effect motion as well. On this reading, when Aristotle says that acting and effecting motion belong to 'another power' he means powers other than those that matter has as a passive subject of change (but nevertheless belong to it as matter). This is why the more dynamic versions of materialism were an advancement over the other kinds: they saw that matter's capacities are not exhausted by those it possesses in virtue of being a *hupokeimenon* but include certain active powers that allows it to generate things. This then sets up Aristotle's third objection that matter is still at best only a proximate agent in the change that is used by the primary agent to bring about its ends.

Both readings have their advantages and drawbacks. Reading A is consistent with the fact that in *GC* I 7 Aristotle himself says that matter qua matter is passive (324b18). In that case we might expect that when he calls something 'matter' in the context of generation he means to identify that thing in its role as a passive subject of change (as what is acted upon and moved). This does not mean that what functions as matter in one context cannot also serve as an agent of change in another. *Meteor.* IV is a case in point. As we have seen *PA* II 1 refers to the four elemental powers (hot, cold, dry, moist) together as the matter of composite bodies. But notice that in *Meteor.* IV Aristotle reserves the term "matter" for the two passive powers (dry/moist), which he contrasts with the two active powers (hot/cold) (e.g. *Meteor.* 378b28–379a3, 379b10–11, b18–19). These two accounts are not incompatible. When Aristotle refers to the hot and the cold as matter in *PA* II 1 he means to pick them out in their role as the ultimate constituents of natural bodies, whereas *Meteor.* IV is focused on their other role as agents in the generation of uniform bodies. Reading A can also accommodate Aristotle's remarks about parthenogenesis in *GA*. Even in those cases where the female generates on her own we can still distinguish the menstrual blood's role as the matter for generation from its role as an efficient cause. In this context the menstrual blood is acted upon and receives form qua matter (in virtue of its of moist and dry elements) but

47 Connell (2001).

acts and imposes form qua efficient cause (in virtue of its own proper heat).[48]

While I am sympathetic to Reading A (e.g. I think the *GA* is working with this conception of matter), admittedly it does not sit comfortably with the doctrine of *Metaphysics* VII 9. There Aristotle is explicit that matter is sometimes capable of setting itself in motion (*kineisthai huph' hautês*). Reading B makes better sense of this. On this reading *Metaphysics* VII 9 is referring to matter, not in its *Physics* I 7 role as passive subject of change, but in its more active role as the efficient cause of that change.

I will not try to settle this here. In the rest of this book I will continue to speak of matter as an efficient cause of generation in cases where I am focused on the causal interactions among the simple bodies and their inherent powers. For example, at *GA* 734b31–3 Aristotle says that the hot and the cold have the ability to make the parts of animals hard and soft, tough and brittle, and so forth. In *GA* II 6 he says that the uniform parts (e.g. flesh, bone, sinew) are all formed by the agency of heating and cooling (743a17–26). Again, at *GA* 739b27 we are told how membranes form around the outside of the embryo as its surface is exposed to heating and cooling. In this case Aristotle goes so far as to allow that these elemental forces bring about change in accordance with what scholars call "material necessity" (*anankaion*, b29). Aristotle attributes this type of necessity to his predecessors at the outset of *Physics* II 8: 'we must speak . . . about necessity and its place in nature, for all writers ascribe things to this cause, arguing that *since* the hot and the cold and the like are of such and such a kind, therefore certain things necessarily are and come to be' (190b10–15). I will leave it open whether or not these elemental forces count as "matter" while operating in their role as efficient causes.[49] While we have seen that Aristotle is sometimes willing to say that matter can effect change (as he does in *Metaphysics* VII 9), by calling something "matter" in the strict sense I think he means to draw attention to its role as the passive subject of a change that gets acted upon and moved (*GC* 324b18).[50]

48 This is consistent with how I analyse the role of menstrual blood in Chapter 5, Section 5.5.2. While I think the evidence of *GA* favours Reading A, this account is compatible with Reading B.

49 Note that they do not count as "material causes" in this role (*Phys.* II 3). Eggs, flour, water, and sugar count as material causes of the cake insofar as the cake is made *out of* them (they are the matter of the cake). As we have seen, Aristotle sometimes refers to the elements as 'the matter of composite bodies' (*PA* 646a13–17, *De juv.* 466a21–3) insofar as all composite bodies are made out of them. The current suggestion is that these same elements (or a subset of them) are also that *by which* a thing comes to be, i.e. efficient causes.

50 Recall that on Reading B it is the concept of subject (*hupokeimenon*) not the concept of matter (*hulê*) that excludes the idea of being an agent of change. On this reading, it is only matter qua subject that

What is important for my purposes is that Aristotle obviously agrees that natural generation can be regarded in some way as an effect of certain elemental forces operating in their role as moving causes. The importance of this will come out more clearly in Chapter 8. As we shall see, this is in no way a concession to his materialist rivals that matter is sufficient to account for natural generation. For Aristotle insists that the simple bodies and their inherent powers are at best proximate causes of the change that impart motion only by being moved by some prior principle whose nature is to produce beauty, order, and goodness. This is the role he assigns to the formal natures of living things, which operate as primary causes of the change (at least in cases where the features in question come to be for the sake of promoting the organism's good).

4.5 Looking Forward

We have now laid out the full hylomorphic model of substantial generation in its most abstract form. According to this model, the generation of a natural substance involves three principles: (1) matter, which is the subject from which the change proceeds; (2) form, which is the end towards which the process advances; and (3) an efficient cause that directs the process towards that form.[51] The next four chapters explore how this abstract model is reflected in Aristotle's theory of biological generation, which is substantial generation par excellence. In Chapters 5 and 6 I examine what I call Aristotle's "reproductive hylomorphism" in the *GA*, which is a more specific application of this abstract model to the generation of living things. In Chapter 7 I consider a puzzle about the efficient cause of animal generation. In *Physics* II 1 Aristotle characterises naturally generated substances as those whose generation is derived from a 'nature'. Most take nature in this sense to be operating as an internal efficient cause. And so we should expect animal generation to be the type of change that is efficiently caused by the

is passive and incapable of effecting change. That belongs to matter in virtue of some 'other power'. Alternatively, we might try to make the doctrine of *Metaphysics* VII 9 fit Reading A by taking *hulê* there not in Aristotle's own technical sense but simply to refer to the raw stuff from which the product is generated. (This reading was suggested to me by David Ebrey.) However, Aristotle does not announce at the outset of VII 9 that by *hulê* he now means matter in some non-technical sense (cf. *Metaph.* 1033b5–6). And so this move may simply beg the question against defenders of Reading B.

51 As mentioned, I take this extended hylomorphic model to apply only to the generation of biological substances. While elemental transformation and the changes that follow upon them (e.g. annual rainfall) have an order-producing cause (the annual rotation of the sun in the ecliptic), this is not the sort of efficient cause that directs the process towards a form as its end.

nature in the embryo itself. However, as we shall see, Aristotle's reproductive hylomorphism identifies the father as an external efficient cause who generates the offspring using his semen as a tool (e.g. *GA* 730b21–4, 734b36–735a4). In that case, it is difficult to see how animal generation qualifies as a natural change at all. The central aim of Chapter 7 is to reconcile these two alternative perspectives on the efficient cause of animal generation. Chapter 8 then turns to Aristotle's account of the process by which a new organism gradually acquires its substantial form and ask how the three principles identified by the extended hylomorphic model (matter, form, and moving cause) interact in the dynamic teleological context of morphogenesis.

CHAPTER 5

Biological Generation: Part One

I now want to turn to the generation of biological substances in the *Generation of Animals*. I have stressed the fact that Aristotle treats living things as substances 'most of all' (*Metaph.* 1034a3–4) and so biological generation should be seen as substantial generation par excellence. This forms the focus of the next two chapters. The central aim of the discussion is to set out in detail what I shall call Aristotle's "reproductive hylomorphism", which is a more specific application of the basic hylomorphic model from the foundational works to the generation of living things.[1] By way of introduction I shall begin with some remarks about the project of the *GA* itself.

5.1 The Project of the *GA*

In what follows I attempt to demonstrate how Aristotle's mature theory of animal generation in the *GA* is driven primarily by a concern to show how the process conforms to the general hylomorphic model and the doctrine of causes.[2] The latter concern is present from the very outset of the *GA*. Aristotle closes the *Parts of Animals* by telling us that after having given an account of the causes of the parts the next task is to study their generation: 'About the parts, then, the cause owing to which each is present in the animals has been stated for each of the animals in turn. These things having been determined next in order is to go through the things concerning their generation' (*PA* IV 14, 697b29–31). The *Generation of Animals* then opens

[1] For various formulations of Aristotle's reproductive hylomorphism see (e.g.) *GA* 729a9–11, 729b12–21, 730a24–30, 732a3–9, 738b18–27, 765b8–15.

[2] Even Aristotle's account of spontaneous generation in *GA* III 11 is driven by an attempt to show how it conforms to the same hylomorphic model used to explain sexual reproduction. There Aristotle says that 'the right method of inquiry' is to ask what it is in such cases that corresponds to the matter 'which in the females is a residual secretion of the animal, potentially such as that from which it came, by imparting motion to which the principle derived from the male perfects the animal' and 'whence comes or what is the moving principle that corresponds to the male' (see 762a36–b28).

with the following programmatic remarks, which are clearly meant to pick up on this:

> Now that we have spoken about the other parts of animals, both in terms of what is common and separately about the parts peculiar to each kind, stating in what way each is due to this sort of cause, I mean the cause "for the sake of which" – for we have set down four causes: both the cause for the sake of which as end, and the definition of the substantial being (these should be taken more or less as the same), and third and fourth the matter and whence comes the origin of the change – and so, having spoken about the other causes (for both the definition and that for the sake of which as end are the same, and the parts are the matter for the animal, the instrumental parts for every animal as a whole, the uniform parts for the instrumental ones, and the so-called elements of bodies for these), it remains to speak about the parts belonging to animals that contribute to generation (about which nothing definite has yet been said) and about the principle of the moving cause. To inquire into this and to inquire into the manner of generation for each thing is, in a way, the same thing. This is precisely why the account has brought them together into one, putting these parts at the end of our account of the parts, and the beginning of our account of generation next to them. (*GA* I 1, 715a1–18)

According to this description the *PA* has taken up the study of animals and their parts from the perspective of their final, formal, and material causes. It remains for the *GA* to study the efficient cause.[3] But this is not all Aristotle says in this passage. We are also told that the program of the *GA* is divided into two main projects: to speak about the parts belonging to animals that contribute to generation, and to inquire into the principle of motion. This latter project is more properly connected with the study of animal generation itself, while the former project (to study 'the parts that jointly contribute to generation by which the female is thought to differ from the

[3] Balme (1992, 127) dismisses this opening passage of the *GA* as spurious and penned by the hand of a misguided later editor. First, he claims that stylistically this introduction is 'so different from the pointed introduction to *PA* I' that it couldn't have been written by Aristotle. Second, he argues that it is false to say that the *PA* only deals with formal, final, and material causes, while the *GA* examines the efficient cause. Balme's first argument is weak, and any claims about sounding "un-Aristotelian" are clearly subjective. Nor does he cite any stylometric evidence to support his view. As to the second point, Balme misses the sense in which the *GA* studies the efficient cause of animals. I discuss this below. For now, we can simply note that this is exactly what we should expect given what Aristotle says about the study of generation at *Ph.* II 7, 198a32–6: 'The question "why", then, is answered by reference to the matter, the form, and the primary moving cause. With respect to coming-to-be, it is mostly in this last way that causes are to be investigated.' Finally, as we shall see, the description of the aims of the *GA* in this passage not only matches what Aristotle actually does, it fulfils a promise made in *PA* and so is exactly what we should have expected him to do here. So, whether or not the passage was actually penned by Aristotle himself (and Balme provides no evidence that it was not), it is an accurate enough reflection of the aims of the treatise to be taken seriously.

male') serves as a transition from the study of the parts in the *PA* to the study of their generation in the *GA*.[4]

Earlier in the *PA* Aristotle had, at several points, postponed the discussion of the parts connected with generation on the grounds that their study more properly belonged to the *GA* (*PA* II 7, 653b9–18; II 9, 655b24–7; IV 4, 678a21–6). Fulfilling this general promise is the goal of *GA* I, which constitutes the first of the two main projects set out above. Now Aristotle divides the generative parts into (a) the uniform parts (seed and milk) and (b) the instrumental parts (uterus, seminal channels, and testicles). The instrumental parts are discussed first in *GA* I 2–16, while the discussion of the uniform parts is taken up in *GA* I 17–22 where Aristotle defends his own theory of seed (*sperma*) against the so-called theory of pangenesis. He draws attention to this division in the closing lines of Chapter 16 just before embarking on the inquiry into the nature of seed: 'These, then, are the instrumental parts related to generation that belong to animals, of which we had not spoken earlier. Concerning the uniform parts, we had not discussed seed and milk' (721a27–30; cf. *PA* II 7, 653b9–18). The account that follows in *GA* I 17–22 also fulfils the promise in *PA* IV 10 to defend the hypothesis that semen and menstrual blood (two forms of animal seed) are both "residues" in Aristotle's technical sense. The claim that semen and menstrual blood are residue was used as a premise in the argument at *PA* 689a4–13. But that premise was simply assumed there (*hupokeisthô*); it will be 'proven' (*deixthêsetai*), Aristotle says, in a later work. That proof comes in the argument at *GA* I 18, 724b24–725a4 whose conclusion states: 'Therefore, it is necessary that seed is a residue.' (The connection with *PA* IV 10 helps to explain, at least in part, why Aristotle spends considerable time in *GA* I trying to show that seed is a residue and not, as others argued, a colliquescence, e.g. *GA* 724b23–725a28.) Finally, the discussion of milk is postponed to *GA* IV 8 since it is produced later in the process of generation, which is in turn explained by the fact that while seed is 'an origin of generation' milk is 'present for the sake of generation' (*PA* II 7, 653b9–18; cf. *GA* II 6, 742a19–b17).

In this way *GA* I 2–22 can be seen not as part of the study of animal generation proper but as a "methodological joint" connecting the study of the *being* of animals in the *PA* to the study of their *coming-to-be* in the *GA*. (An anatomical joint is said to be 'the beginning of one thing and the end of another', *MA* 702a22–3.) This is what Aristotle means at the end of *GA*

[4] I originally presented this interpretation at the 5th Pittsburgh–London Workshop on the *Generation of Animals* at the University of Western Ontario (25–27 May, 2010).

715a1–18 when he says: 'This is precisely why the account has brought them together into one, putting these parts [sc. those connected with generation] at the end of our account of the parts [*PA*] and putting the beginning of our account of generation [*GA*] next to them.'[5] This does not rule out the possibility that Book I plays some additional role in the broader structure of the *GA*. As we shall see, the discussion of seed in *GA* I 18–22 also serves the function of supplying proof for the hylomorphic model that provides the framework for Aristotle's study of animal generation in the rest of *GA*.

The second project that Aristotle announces in the opening passage of the *GA* is to inquire into the nature of the efficient cause of animals and their parts. It is this latter project that more properly constitutes the study of animal generation itself. And this is exactly what we should expect given what Aristotle says about the study of generation in *Physics* II 7, 198a32–6: 'The question why, then, is answered by reference to the matter, the form, and the primary moving cause. For in respect of coming-to-be, it is mostly in this last way that causes are investigated: What comes into being after what? What was the primary agent and patient? And so on at each step in the series' (198a32–6, Hardie and Gay translation). This passage gives some clue as to what Aristotle means at the end of our *GA* passage by saying that the inquiry into the efficient cause of an animal and the inquiry into its formation are 'in a way' one and the same thing. For when it comes to

[5] This way of reading the *GA* differs markedly from the view proposed by Balme (1992, 127). Balme sees Book I as 'a theoretical discussion from first to last' but claims that 'its thread has been obscured by some ancient editor who believed it to be descriptive anatomy' (see note 3 above). He argues that Book I aims to set out Aristotle's 'general theory of reproduction in a long but connected argument which continues to the end of II.3', while the rest of the *GA* goes on to offer 'descriptive' accounts of development 'genus by genus'. On this reading, there is a significant break between the 'dialectical' move *towards* first principles in *GA* I–II.3 and the rest of the treatise which he claims proceeds *from* first principles. The main problem with this reading is that it ignores Aristotle's explicit remarks about the structure of the treatise (e.g.) at *GA* 715a1–18 (I have already commented on this above). As I have shown, Book I is making good on certain promises in the *PA* to take up the generative parts that Aristotle had explicitly postponed to the study of generation (*PA* 653b9–18, 655b24–7, 678a21–6). On Balme's reading these connections are lost. Moreover, the discussion of the generative parts in Book I doesn't fit neatly into Balme's picture. *GA* I 2–16 contains *PA*-style explanations of the instrumental parts connected with generation. And there is hardly anything in those chapters that would count as a dialectical argument that 'moves from agreed starting points to uncover more basic principles'. So Balme's reading would at least have to be modified so that *GA* I 2–16 constituted the promised discussion of the generative parts, while the so-called long connected argument in which Aristotle sets out his general theory of reproduction begins later in *GA* I 17. But in fact, Aristotle's remarks at the end of *GA* I 16, which look forward to the next step in the account, clearly show that he takes the whole of Book I to be part of the same project (721a27–30). Finally, the part that is supposed to be purely "descriptive" itself contains a considerable amount of theory (e.g. *GA* IV 1–3, V 1, 8). In general, Balme's picture of the structure of *GA* does not work. Nowhere does Aristotle self-consciously organise the *GA* into a preliminary "theoretical" part on the way *to* first principles and a subsequent "descriptive" part that moves *from* first principles.

generation the inquiry is focused on questions like "What comes into being after what?" and "What is the primary agent and patient?", and this just is to investigate the process by reference to the moving cause.

I will return to the question of the efficient cause of generation in Chapter 7. As we shall see, there is a puzzle here about whether the generation of an animal, properly speaking, has an internal or external efficient cause and what this tells us about Aristotle's definition of nature in *Physics* II 1. In this rest of this chapter and in Chapter 6 I shall focus exclusively on the other main goal of the *GA*, namely, to defend the hylomorphic model as the best means of explaining the facts surrounding animal generation.[6]

5.2 Reproductive Hylomorphism

What I shall call Aristotle's reproductive hylomorphism can be analysed into three main claims:

(A) The generation of a living thing involves a division of reproductive labour in which male and female principles[7] make different, yet complementary, contributions to the thing that comes to be.
(B) Within this division of labour the male is the efficient cause that supplies the form, while the female provides the matter that receives that form.
(C) In the process of becoming a new substance, the matter that forms the starting point of the change is gradually transformed from a prior state of potentiality to a more mature state of actuality as a result of being acted upon by another substance that already bears the form in a state of completion.

Each of these propositions raises several questions that will occupy the rest of this book.

Proposition (A) reflects an important feature of Aristotle's more general hylomorphic model. As we have seen, *Physics* I 7 argues that all coming-to-be proceeds from matter and form and that these two principles play different causal roles in the generation of a substance (e.g. *Ph.* 190b10–23). The idea

6 For places where Aristotle explicitly invokes the hylomorphic model as the best explanation for some aspect of generation see, e.g., 727b14–16, 728b32–729a21, 730a24–b8.

7 Aristotle routinely distinguishes the male and female principles, which he identifies with certain capacities or powers (*dunameis*), from the male and female animals that house those principles. This distinction allows him to hold that even plants contain male and female *principles*, even though they are not divided into male and female individuals.

that biological generation involves a similar division of labour was itself an innovation on Aristotle's part.[8] As we shall see, some of his predecessors denied that both parents make a reproductively significant contribution to generation, while others agreed that both parents make a contribution but denied that they make different contributions. What grounds, then, does Aristotle have for thinking generation involves a reproductive division of labour at all let alone one drawn along hylomorphic lines?

A second set of questions surrounds the content of the model itself (Proposition B). What does Aristotle mean by saying the male contributes "the form" while the female contributes "the matter"? What exactly does that thesis entail? At *GA* 738b21–3 Aristotle says that these are the defining contributions of males and females: 'for we say that this is the capacity that each has and what it is for something to be male or female'. Does he also mean to say that these are their *only* contributions? If instead (as I shall argue) the female also provides part of the offspring's form insofar as she provides capacities of its soul, what does this imply about the unity of the animal form? Finally, Aristotle holds that plants contain male and female principles even though in most cases they are not separated into distinct sexes.[9] So why, in the case of animals, do these sexes exist in separate individuals? Aristotle himself takes the case of plants to show that it is not necessary for generation that these two principles exist separately. Instead he argues that in animals the principles are separated because it is better to keep them apart. In what sense is it "better" to house these principles in distinct bodies? I postpone this question until Chapter 9.

Finally, Proposition (C) raises several questions of its own. What exactly is the efficient cause that is responsible for transforming the matter into a new substantial individual that bears the same form as its generating parents? Proposition (B) suggests that this role is played by the father, which is external to the developing thing. But *Physics* II 1 characterises natural changes as those whose moving principle is *in* the thing that undergoes the change, which suggests that what transforms the matter during its development is a principle in the developing thing itself (the embryo's own nature). How do we reconcile these two perspectives on the efficient cause? Further, how do the formal nature and material nature of a living thing interact in the process? And what role does so-called

[8] Even if Kelsey (2010, 109–10) is right that hylomorphism itself was not an innovation on Aristotle's part, it is clear from the *GA* that Aristotle does not think any of his predecessors saw hylomorphism as the best way to explain the facts about generation.

[9] Pace Kosman (2010, 147).

material necessity play in this? I take these questions up in Chapters 7 and 8.

I begin here with a brief overview of the hylomorphism of animal generation, which is first introduced at the outset of *GA* I 2:

> Concerning the generation of animals, we must discuss various questions case by case in the order that they arise, and we must connect our discussion to what has been said. For, as we say, one could above all set down the male and female as principles of generation, the male as the principle of the change and generation, the female the matter. One will be most persuaded of this by considering how and from where seed comes into being. For, although the things that are naturally generated are formed from this, we must not fail to notice how this comes to be from the female and the male. For it is because this sort of part is secreted from the two sexes (the secretion taking place in them and from them) that they are principles of generation. (*GA* I 2, 716a2–13)

In *GA* I 1 Aristotle had noted that some animals come to be from the union of male and female (715a17–29) which makes them *archai* of generation in the sense of starting points or beginnings of the change. But male and female are also *archai* in the sense of causes: the male as the efficient cause, the female as supplying the material cause. That male and female are *archai* in these two senses is not something that has been said before. Instead it is the hypothesis to be established in the treatise that follows.[10]

Aristotle says that we will be 'most persuaded' (*malista pisteuseie*) of this hypothesis by attending to the facts about seed.[11] This the aim of *GA* I 17–22. That investigation takes as its starting point the pre-scientific concept of seed as the sort of thing *from which* naturally generated organisms first come to be (724a17–20). Aristotle then goes on to argue that seed must function either as the material cause of generation or as the efficient cause, since these correspond to the two senses of "from which" (*ek*) that are most appropriate in

[10] While part of the project of the *GA* is to establish the thesis that male and female are principles in the sense of causes, as we shall see in Chapter 9 they are not *causally basic* principles. For Aristotle thinks that their existence can itself be traced to a 'higher principle', namely, the divine.

[11] At this point it is worth remembering that Aristotle's word *sperma* carries several different meanings (see Chapter 2, note 6). In its widest sense it refers to the reproductive material produced by any sexual organism (its "seed"). Aristotle counts menstrual fluid as a form of *sperma* in this sense (e.g. 728a26–7, b23, 750b4–5, 767b16–17, 771b20, b22–3), which is the analogue of semen in males. Indeed, as he says at the end of the *GA* I 2 passage, it is because the female generates *sperma* that she counts as an *archê* of generation (716a11–13). Admittedly, there are some passages where Aristotle seems to deny that females produce *sperma* (e.g. 727a26–30), but those passages are best read by taking *sperma* there in the narrow sense to refer to male semen. What Aristotle means to deny in those passages is that females produce *functionally the same kind* of seed as the male 'as some allege' (e.g. 727b6–8, 728a27–31). In other words, he means to deny that females produce semen in addition to menstrual blood (the so-called twin semen theory).

this case: 'One thing might come to be "from" another in many ways . . . It is clear, though, that seed (*sperma*) falls under one or the other of two senses: what comes to be is formed "from" it either as "from matter" (*hôs ek hulês*) or as "from the primary moving cause" (*hôs ek prôtou kinêsantos*)' (724a18–b7). Since (as Aristotle will show) both males and females contribute seed, and since these make different causal contributions to generation, the job of *GA* I 19–22 becomes sorting out whose spermatic contribution plays which role in the process. Whose seed makes its contribution to generation 'as matter and that which is acted upon' (*hôs hulên kai paschon*) and whose makes its contribution 'as form and that which acts' (*hôs eidos ti kai poioun*) (*GA* I 18, 724b4–6)? According to the model defended in *GA* Book I animals come into being from female seed in the first sense and male seed in the second.

5.3 Division of Labour

At the core of this model is the claim that generation involves a division of labour in which male and female principles make different, yet complementary, contributions to the thing that comes to be from them (Proposition A). There are two ways that one might deny this. Androcentric theories of reproduction deny that the mother makes any causal contribution to generation at all. All she provides is the place for the male's seed to develop (cf. 726b1–2). This model is famously represented in a passage from Aeschylus' *Eumenides*: 'She who is called the mother of the child is not a parent, but a nurse of the newly sown embryo. The one who mounts is the parent, and she but a stranger, a friend, who, if fate spares his plant, preserves it till it puts forth.'[12] Androcentric theories like this are a non-starter for Aristotle.[13] The empirical evidence clearly shows that menstrual blood is a spermatic residue that is the analogue of semen in males (*GA* I 19, 726b30–727a4; cf. *GA* IV 1, 765b19–35, 766a30–b3), which makes her a principle of generation along with the male (716a2–13). More importantly, androcentric theories cannot explain the fact that offspring often resemble their mother and her ancestors, which Aristotle thinks an adequate theory of heredity must explain (*GA* IV 3, 769a16–26).

[12] *Eumenides* 658–9 (quoted from Mayhew 2004, 28 with modification). For discussion see Tress (1992, 319) and Mayhew (2004, 28–9).

[13] Interestingly, Aristotle does take seriously at least the possibility of a gynocentric model where the mother generates the offspring all by herself without any input from the father. This is the *aporia* that drives the discussion in *GA* II 5 (discussed in Chapter 6).

Alternatively, one might accept that both parents make a reproductively significant contribution to generation but deny that they make different contributions. According to the so-called theory of pangenesis, for example, both parents make an identical contribution to generation, specifically, they both contribute semen.[14] On this theory seminal fluid is made up of tiny bits of tissue drawn directly from each part of the parent's body. At conception the semen from the male mixes with the semen from the female forming a single seed from which the new individual will eventually grow. This is supposed to explain resemblance: as development unfolds the seminal materials are separated out again and redistributed to the same places whence they were originally drawn according to a principle of like-by-like, in effect reconstituting the parent's body in the offspring (see *GA* II 4, 740b12–18). Aristotle spends considerable time in *GA* trying to refute this rival theory (see especially *GA* I 17). The reason he is worried about pangenesis, I submit, is that he sees it as a threat to the causal division of labour at the core of his own hylomorphic model of generation. First, the twin semen hypothesis entails that male and female play the same causal role in generation. Like every other part of an animal, seed is defined by its function. If both males and females contribute the same kind of seed (viz. semen), it would follow that both play the same causal role in generation. Second, the pangenesis aspect of the model entails that they both play the role of *material* cause. For on that theory living things breed true to type because the offspring's body is literally made out of (numerically) the same matter drawn from the parts of the parent's body. (This aspect of the theory is Aristotle's target in *GA* I 21.) Both of these claims threaten to undermine the division of labour at the core of Aristotle's own reproductive hylomorphism.

Given all of this we might be surprised that Aristotle does not explicitly defend the claim that generation involves a division of labour within the *GA* itself. Instead many of his arguments take that hypothesis for granted. Consider the way Aristotle frames the inquiry into the nature of seed in *GA* I 18. He begins the investigation from the pre-scientific concept of seed as

14 This, I take it, is what Aristotle has in mind at the start of *GA* I 17 when he asks 'whether the female also contributes any semen or not, and if not semen, whether she contributes nothing else [to generation] or whether she contributes something other than semen?' (721a35–b2; cf. 726a31–b2). Aristotle of course denies that the female emits semen 'as some allege' (727b6–8). Peck and Connell routinely use "semen" as a translation for female *sperma*, which obscures this crucial point. The English word "semen" refers to the milky fluid that males emit which contains spermatozoa. Aristotle reserves the word *gonê* for this (though in some contexts he uses the more generic *sperma*). All parties agree that semen is emitted by males. What 'some allege' is that females *also* emit semen *in addition* to menstrual blood. This is what Aristotle's theory of female seed is meant to correct.

'the sort of thing *from which* naturally generated organisms first come to be' (724a17–20; cf. 721b6–8) and then attempts to analyse the contribution that seed makes to generation by considering the different ways in which one thing comes to be "from" another. He concludes that the two most relevant senses pick out the material and efficient cause, respectively (724a20–b6). Aristotle's account here does not begin from commonly held beliefs (*endoxa*) that presume no technical information or theory (pace Bolton 1987). Instead, the basic parameters of the inquiry are clearly shaped from the start by a prior commitment to the hylomorphic model of generation, specifically, to the idea that generation involves a causal division of labour. For it obviously influences Aristotle's choice of which senses of *ek* are most appropriate to the study of biological generation: he assumes from the start that the way to study the phenomenon is to identify which parent's seed is a cause of the offspring as matter and which is a cause as form; and that in turn presupposes that there is a reproductive division of labour involved.

While it is true that Aristotle never defends this more basic assumption within the *Generation of Animals* itself, it does not follow that he helps himself to this hypothesis. (That would certainly beg the question against his rivals.) Consider *GA* 72924–34, where the principle is used as a premise in one of Aristotle's key arguments for reproductive hylomorphism:

> (1) It is necessary for there to be that which generates and that out of which the product is generated. (2) Even if these are one, they must at least differ in form and the account of each must be different. (3) In those living things in which the two capacities are separate, the body and substance of the agent and the patient must be different. (4) If, then, (a) male is mover and agent (*kinoun kai poioun*) while (b) female is the patient that gets acted upon (*pathêtikon*), then (5) what the female contributes to the semen of the male is not semen but matter. (6) And this is exactly what corresponds to the phenomenon; for the nature of the menstrual fluid is in accordance with first matter. (*GA* I 19, 729a24–34)

Premises (1) and (2) clearly form part of the theoretical ground for claiming that reproduction involves a division of labour. For together they state that all coming-to-be requires an agent (that which generates) and a patient (that out of which it generates) and that these two principles must be at least separate *in account* insofar as their functions are different (cf. 765b8–15). *A* and *B* are separate in form and account if the definition stating what it is to be *A* does not make reference to *B* or vice versa such that *A* and *B* can be defined independently of one another.[15] Aristotle has in mind here the

[15] Corcilius and Gregoric (2010).

difference between animals and plants. In animals the male and female principles are separate in the stronger sense of being separate in number and magnitude (i.e. they are housed in different bodies). In plants the male and female principles are united in the same individual, but we can still make a conceptual distinction between the active (male) principle and the passive (female) principle (see *GA* I 23, 730b32–731a4). Even here, then, the two principles are still separate in account. While (1) shows that biological generation must involve two separate causal principles, this only commits Aristotle to the weaker claim that these principles must be separate in account. It does not entail the stronger claim (3) that they must be housed in separate bodies (i.e. separate in number, magnitude, or place). All Aristotle says here is *that* these principles are separate in this stronger sense in the case of animals, but he does not tell us *why* they are. That argument will have to wait until *GA* I 23–II 1 (see Chapter 9).

Just before this passage Aristotle characterises the argument as one that derives its conclusion 'according to reason' from premises that are sufficiently 'universal' (*kata ton logon katholou skopoumenois*; cf. *GA* 729b9–10: *katholou te gar episkopousin*), which he contrasts with those that are established on the basis of perception (e.g. 740a3–5). By referring to these premises as universal Aristotle at least means that the principles in question extend beyond the specific case of animal generation to some wider class of phenomena.[16] The universal principles in this case are (1) and (2), which apply to all cases of coming-to-be. If this is right, then we might expect them to be established in some other treatise dealing with the broader science of nature. And I think we find that argument in *Physics* VIII 4–5. There Aristotle argues that in any change that which moves is always distinct from that which is moved. Even in the case of self-moving systems, the system moves itself as a whole in virtue of one part that effects motion (which is itself unmoved) and another part that undergoes motion.[17] From this general principle the above argument takes aim at the twin semen hypothesis by showing that what the female contributes to generation cannot be semen, since in that case there would be two agents. (That semen is the agent in generation was likely a shared assumption. And it is obvious that males at least emit semen.) Therefore, since generation requires an agent *and* a patient, and since we know that males emit semen,

[16] For Aristotle's epistemological concept of a universal see *APo.* I 4.

[17] See *Ph.* VIII 5, e.g. 254b29–257b15. Plants might be considered self-moving reproductive systems. Even in this case one part moves (the male principle) while another part is moved by it (the female principle).

the female's spermatic contribution must be playing the role of matter (since matter qua matter is passive, *GC* 324b18).

So Aristotle does not beg the question against the rival theory by simply assuming that biological generation involves a causal division of labour, even if the hypothesis is taken for granted in the *GA* itself. That claim is grounded in the more general principle that in all cases of generation it is necessary for there to be that which generates and that out of which the product is generated and that, even if these are housed in the same individual, the two capacities must at least differ in account. The proof for this theorem, I have suggested, is to be found in the arguments concerning self-motion in the *Physics*, which is the part that deals with the wider science of nature. It follows from this that since males do not provide matter for generation, then if the male makes any contribution at all, it must be the principle that imposes form on the matter coming from the female (this is the main upshot of *GA* I 21).

5.4 Two Applications of the Hylomorphic Model

In the exceptional preface to his translation of the *GA*, Peck begins his analysis of Aristotle's reproductive hylomorphism with the simple question: What exactly is meant by form and matter in this connection? One popular way of answering this question is to read them directly off of the analogy with craftsmanship. Just as the sculptor makes a statue of Theaetetus by imposing a certain shape and form on the bronze, so too the father (using his semen as a tool) moulds the materials provided by the female into a new individual.[18] While Aristotle certainly makes extensive use of the craft analogy in the *GA*, this way of approaching the model is of limited value. Aristotle uses the craft analogy mainly to explain how the father makes his contribution to generation because he thinks there are certain features of craftsmanship that are well-suited to illustrate his account. Two features in particular are crucial for Aristotle's reproductive hylomorphism. First, the craftsman does not contribute anything material to the product. For example, the carpenter does not contribute his own flesh and bone as matter for the house; instead, he transmits a form without matter. Second, in typical cases of craftsmanship the form is transmitted through the movements of his tools (730b8–19) which are said to embody the activity of his craft (734b36–735a4, 740b28). Aristotle thinks that both of these features characterise the male's contribution to generation: males

[18] See Cooper (1988, e.g. 30–1), Furth (1988, 119), Mayhew (2004, 44).

do not contribute any matter to the offspring but only form; and the way they transmit form is through "motions" in the semen, which serves as his "tool" (e.g. *GA* I 21, 730b19–23).

But it is important not to read too much into the analogy. Aristotle is not suggesting that the analogy with human craftsmanship is so close that we can think of the father as giving form to the menstrual blood by literally shaping the parts of the offspring out of it like a sculptor using a chisel.[19] Even when Aristotle compares embryogenesis in insects to modelling in clay (730b24–32) the point of analogy is simply the fact that in both cases the materials are brought to the agent so that he acts directly on the matter rather than indirectly using tools. In these cases, Aristotle observes, the female inserts her ovipositor into the male and brings the materials directly up to the region around his heart where the source of natural heat resides (730b24–32, 738b10–18). But nothing in the text requires that we take the analogy so literally as to suppose that there is something inside the male that physically moulds the offspring's body out of those materials.

In rejecting the literal reading of the craft analogy, I also mean to reject the purely morphological reading of form (*eidos*) that it implies. When Aristotle says the male alone provides the animal *eidos*, he does not mean he is solely responsible for the outward physical appearance of the completed offspring's body (*eidos* in the pre-theoretical sense of *morphê*). That is not the form specified by the hylomorphic model of animal generation.

What I will argue is that Aristotle's reproductive hylomorphism is best understood as providing a general explanatory framework which he applies to different aspects of generation.[20] In its most generic formulation the model says that the male provides the form while the female provides the matter that receives that form. However, form and matter here cannot be specified independently of the particular explanatory context in which that model is being deployed (e.g. we cannot understand those terms simply by reading them off the analogy with craftsmanship). Instead what Aristotle means by form and matter depends on which aspect of generation is under consideration. As we shall see, the *GA* deploys the hylomorphic model in connection with two different aspects of generation.[21] In Book I he uses it

[19] Connell (2016, 121) argues that the craft analogy is even more limited when it comes to understanding female matter. For her material contribution is not an inert lump of unorganised stuff like the bronze or marble. I return to this below.

[20] Here I follow Witt (1998, 121) in treating Aristotle's hylomorphism, not only as a metaphysical thesis about the being (or ontological make-up) of sensible substances, but also as a model for explaining natural phenomena including the generation of those substances.

[21] I am in general agreement with Kosman (2010, e.g. 149) here. However, our interpretations disagree considerably over how we fill out these two versions of the model.

primarily as a way to analyse the causal contributions that male and female seed make to the initial formation of the embryo, which (I shall argue) functions as the pre-existing subject from which generation proceeds (the *terminus a quo*). In this context semen "gives form" to the embryo only in a purely mechanical sense of imposing a limit or boundary on the relatively indeterminate fluid supplied by the female – a process that Aristotle routinely compares to the coagulation of milk. In Book II Aristotle uses the model to explain the relative contributions that the mother and father themselves make to the substance that results from the change. Here the form supplied by the father is identified with the offspring's soul, while the mother is said to provide the body that serves as the underlying subject of that form. However, as we shall see, it is not the whole soul but only the sensory part of the soul that constitutes the father's *exclusive* contribution to the generation of an animal (*GA* II 5). This is the *eidos* in the strict sense specified in the definition of what it is to be an animal (the possession of this form is what makes a living thing an animal).

On this interpretation the *GA* yields two substantive applications of the hylomorphic model of generation. The first is an analysis of the causal contributions that male and female seed make to the formation of the embryo (embryogenesis), while the second version is an analysis of the contributions that the mother and father make to substance that comes to be from it (the mature offspring). As we shall see, what counts as "matter" and "form" is different in each of these two contexts. I take up the first version in the remainder of this chapter, while Chapter 6 examines the second version.

5.5 Embryogenesis

5.5.1 Male Form

Here is a sample of passages where Aristotle invokes what I have identified as the first version of his reproductive hylomorphism, which is the version that dominates Book I:

> The male provides the form and starting point of motion, while the female provides the body and the matter (*to men arren parechetai to te eidos kai tên archên tês kinêseôs, to de thêlu to sôma kai tên hulên*). (*GA* I 20, 729a9–11)

> The female qua female is the patient that is acted upon (*pathêtikon*), while the male qua male is the agent and whence comes the starting point of the change (*poiêtikon kai hothen hê archê tês kinêseôs*). Therefore, taking the most

> general formulation of each (qua agent and mover and qua patient and moved) we can see that one thing comes into being from male and female only in the sense that a bed is formed from the carpenter and the wood or a ball from the wax and the form. It is clear, then, that it is not necessary for anything to be discharged from the male, and if he does discharge something what comes into being does not come to be from it as a constituent part (*hôs enuparchontos*) [i.e. as matter]. Instead it comes to be from semen as a thing comes to be from the mover and the form (*hôs ek kinêsantos kai tou eidous*) just as health comes from the medical art. (*GA* I 21, 729b12–21)

> What has been said makes it clear that in the case of animals that emit semen the semen is not drawn from the whole body and also that in generation the contribution of the female to the things that are constituted from it (*tois sunistamenois*) is different from that of the male, but that the male contributes the starting point of motion while the female supplies the matter. This is why the female cannot generate all by herself for there must be a principle that moves the matter and defines it (*archês kai tou kinêsontos kai dioriountos*). (*GA* I 21, 730a24–30)

In these passages Aristotle is talking primarily about embryogenesis where the semen acts on the menstrual fluid and forms it into an embryo.[22] In this context the semen "gives form" to the embryo in the purely mechanical sense of imposing a limit or boundary (*dioriountos*) on the relatively indeterminate fluid causing it to set.[23] There are two reasons for adopting this deflationary reading of form in these passages.

First, it is fairly certain that in *GA* I 18–22 (the main discussion of the hylomorphic model in Book I) Aristotle is focused almost exclusively on the initial formation of the embryo. He does not even raise the question of what it is that makes the parts of the developing animal until *GA* II 1 (733b31–2). And the account of morphogenesis itself does not begin in earnest until as late as *GA* II 4, 739b33, which begins with the line: 'Once the embryo (*kuêma*) has now been set'. Since Aristotle is not yet concerned with how the embryo develops its mature structures in *GA* I 18–22, but only with how it is initially formed, when he says at that early stage that the male provides form we should not expect "form" to pick out the animal's mature substantial form but rather some feature of the primitive embryo.

Second, the deflationary reading makes good sense of the central analogy Aristotle deploys in *GA* I 18–22 to explain *how* semen imposes form on

[22] In *HA* VI 3 Aristotle describes the product of this event as a 'speck of blood' that 'beats and moves as though alive' (*HA* 561a6–18).

[23] For the embryological significance of the verb *sunistanai*/*sunistasthai* ('set') see Peck (1942, lxi).

menstrual blood, namely, the interaction between rennet (or fig juice) and milk:

> The male provides the form and the starting point of motion while the female provides the body and the matter, just as in the coagulation of the milk where the milk is the body and the fig juice or rennet contains the principle that causes it to set (*to tên archên echon tên sunistasan*). (*GA* I 20, 729a9–11)

> When the discharge in the female's uterus is fused together by the semen of the male it is virtually the same sort of thing as the action of rennet on milk. For rennet is milk that contains vital heat, which draws the bits of material together and forms (*sunistêsi*) them into a single mass. The nature of milk and menstrual blood is the same, and the action of semen on the menstrual blood is the same as rennet on milk. (*GA* II 4, 739b20–6)

Aristotle must be comparing the action of rennet (and fig juice) on milk to the causal interaction between semen and menstrual blood in the formation of the early embryo. For the analogy would be completely inadequate to explain the acquisition of form in any more robust sense than that. For example, if by form here Aristotle had in mind anything as sophisticated as capacities of soul, then the analogy would leave unexplained precisely what it was supposed to explain since curds do not acquire soul as a result of coagulation. By contrast the use of this analogy is fully adequate to explain the phenomenon, if "giving form" in this context is understood in a strictly mechanical sense of imposing a limit or boundary on the relatively indeterminate menstrual blood.

The deflationary reading also has the virtue of rendering Aristotle's use of this analogy less naive than it may otherwise appear. It would certainly be naive to think that something as complex as animal generation could be understood in such simple terms as semen "curdling" menses. Such an account would be, as Furth (1988, 119) puts it, quite crude and childlike by present-day lights compared to the actual mechanisms involved. But if all Aristotle means to convey with this analogy is how semen forms the primitive embryo (embryogenesis), then his account is not overly simplistic at all. In fact, it is an example of some rather sophisticated scientific reasoning. Aristotle had no way of directly observing the actual causal interactions involved in embryogenesis. While it is true that he recorded the stages of avian development by cracking open eggs at different time intervals (*HA* VI 3), at best those observations told him *that* development occurs in this way; they would not have revealed anything about the actual causal mechanisms involved. Therefore, in order for an empirically driven scientist like Aristotle to establish that his own proposed causes of embryogenesis were adequate to produce the effect in question, he would have to resort to other argument strategies besides direct observation. In this particular case Aristotle appeals to a form of analogical reasoning that begins from what is more familiar to us.

Scientific arguments from analogy typically work by studying the dynamics of some observable phenomenon – the source domain *A* – and then extrapolating from this to the phenomenon under investigation – the target domain *B*. If the analogy between *A* and *B* is sufficiently close, then we are justified in drawing conclusions about what happens in *B* based on what we observe happening in the more familiar case *A*. More specifically, reasoning by analogy allows us to make judgements about how the unobserved causes operate in *B* based on our observations of the causes at work in *A*. One of Aristotle's typical source domains for understanding the biochemical mechanisms at work in the living body is cooking (see especially *Meteor.* IV). For cooking also relies on producing certain changes in biological materials through chemical reactions involving heating and cooling. Aristotle's model of choice for understanding the causal mechanisms involved in embryogenesis (or at least fertilisation) was the coagulation of milk. Experience shows that if you store milk in a bag made from the stomach of a young ruminant the rennet will transform the milk into solid curds and liquid whey. The Greeks figured out that the same effect could be produced by other sources like fig juice.[24] It is this process that Aristotle thinks is analogous to the way that semen informs the menstrual blood: menstrual blood is the analogue of the milk, while semen is the analogue of the rennet insofar as both possess heat (their active ingredient).[25] Given the close analogy Aristotle sees between the coagulation of milk and the formation of an embryo, it was reasonable for him to assume that the same underlying causal mechanisms must be at work in both processes. Based on this reasoning he concluded that when the father's semen comes into contact with the menstrual fluid the heat in the semen initiates a chemical reaction that causes the small bits of spermatic residue present in the menstrual fluid (729a23) to fuse together into a unified mass. In both cases the chemical reactions involved produce an

[24] Fankhauser (2017).

[25] Although Aristotle refers to this as *zôtikê thermotês* (living heat or vital heat), we should not take this as an endorsement of vitalism. He is not suggesting that semen has some kind of special heat that gives it the power to infuse the embryo with life. Instead, by referring to this heat as *zôtikê* Aristotle is making the ordinary, though non-trivial, point that it is the sort of heat that is generated naturally by a living thing (by various internal metabolic processes) as opposed to, say, the heat produced from fire. (This is all he means in the later passage at *GA* II 3, 736b29–737a7.) For a more vitalistic take on this concept see Freudenthal (1999).

unorganised, amorphous mass: an embryo in the one case, a curd in the other (724b18).[26]

5.5.2 *Female Matter*

I have argued that in *GA* I Aristotle applies the hylomorphic analysis to the formation of the embryo, which *Physics* I 7 identifies as the subject from which the generation of a living thing proceeds. According to this version of the model, semen forms the embryo by imposing a limit or boundary on the fluid supplied by the female and then catalyses its development. This is what Aristotle means in Book I by saying the male provides 'the form and the starting point of the change'.

Many scholars have taken Aristotle's claim that the female contributes the matter for generation as saying that her contribution is something entirely passive and inert and therefore rather insignificant to the process.[27] Aristotle's use of the craft analogy to articulate his reproductive hylomorphism makes this interpretation particularly tempting. For example, at *GA* 729b12–21 (translated above) Aristotle explicitly compares the female's contribution to wood and wax while the male is associated with the craftsman who imposes form on those materials. This certainly invites the idea that Aristotle views menstrual blood as an inert, unorganised lump of stuff that contributes little more to the generation of a new substance than a shapeless lump of bronze contributes to the formation of a statue. The only active contribution it makes to the finished character of the offspring (on this view) is through its recalcitrance, its ability to resist being informed by the organising capacities of the male semen which results in something not-quite-true-to-type like the sculptor who is forced to work with defective bronze.[28] Finally, Aristotle holds that female seed is cold relative to male seed. And in Aristotle's biochemistry heat is the active ingredient through which the important process of concoction is effected while cold is sometimes associated with passivity and the absence of heat (e.g. *Meteor.* 382a37–9).

Other scholars have argued that the *GA* does not conceive of female matter as completely inert but as a highly complex and dynamic factor in

[26] At *GA* 771b22–4 Aristotle recognises a disanalogy between the two cases. But the point of disanalogy is not that semen forms complex functional structures while rennet and fig juice only produce amorphous curds. Rather Aristotle notes that semen has the ability to fuse together separate bits of menstrual fluid into *multiple* embryos of a species-specific size – as happens in animals that produce many young at once – while rennet and fig juice always produce one big mass (see *GA* 772a22–5). Whatever the significance of this may be, Aristotle is still clearly talking about semen's role in embryogenesis.

[27] For references see Connell (2016, chapter 4). Connell herself rejects this claim.

[28] E.g. Cooper (1988), Cook (1996).

the process of generation. The theory of heredity in *GA* IV 3 clearly means to assign the female's contribution an active role in the production of those family resemblances that make the offspring look like her and her ancestors.[29] And Aristotle's recognition of parthenogenesis shows that in some cases he thinks that menstrual blood is capable of forming itself into an embryo and catalysing its own development in the absence of any contribution from the male (e.g. 730a30–2). In these cases, menstrual blood is both that *out of which* and that *by means of which* the new animal comes into being. This suggests that Aristotle is not thinking of female matter as completely passive and always in need of something else to set it in motion.[30]

While I think this second interpretation is right, it needs to be qualified by making a distinction between female seed (menstrual blood) and its role as the matter for generation. However true it may be that menstrual fluid is something complex and dynamic, by calling it "matter" Aristotle means to identify it with a certain role in the process of generation, namely as the passive subject of change that is potentially the substance that comes to be from it. For matter, as a principle of coming-to-be, is something passive and thus requires another principle to set it in motion (*GC* II 9, 335b29–33).[31] For example, in *GA* I 21 Aristotle argues that semen acts as a cause of the embryo, not by supplying matter that mixes with the material coming from the female, but in virtue of a certain active power that makes it the agent (*he poiousa*) in the change. He contrasts this with the role of menstrual blood, which is the thing that 'is set and receives form' (*to de sunistamenon kai lambanon tên morphên*). It is clear from the remainder of the argument (729b12–21) that by calling

[29] See Henry (2006a, 2006b), Connell (2016, chapter 9). For an alternative attempt to explain maternal resemblance in terms of the active powers of male semen alone see Cooper (1988).

[30] Connell (2016, chapters 3–4). See also Mayhew (2004, chapter 3). A large part of Connell's project is to combat readings of the *GA* according to which the female's matter 'is empty and completely malleable with no specifications of its own' (2016, 319). The main differences between our readings stem from the way we understand Aristotle's concept of matter. On Connell's reading, Aristotle thinks the female only contributes matter and so she needs to conceptualise female matter as something complex and dynamic enough to account for her other roles in generation (including her ability to transmit resemblances). By contrast, I take the *GA* to be working with the concept of matter as entirely passive. But this doesn't mean Aristotle thinks menstrual blood *only* plays the role of matter in generation.

[31] This way of understanding the *GA* theory is compatible with at least one way of reading *GC* II 9 presented at the end of Chapter 4. Recall that at *GC* 335b29–33 Aristotle objects to materialism on the grounds that that 'matter is what is acted upon and moved, while effecting motion and acting belongs to another power'. On one reading, Aristotle thinks matter qua matter is characterised entirely by its passive capacity to be acted upon and moved. However, as I go on to argue in Chapter 6, Aristotle does not think menstrual blood's role is confined to serving as the matter for generation. It also conveys motions that transmit part of the offspring's form. In this role menstrual blood is no longer functioning as matter but as an agent of change.

menstrual blood matter in this context, Aristotle means to contrast its passive role in generation with the active role of semen as the cause that imposes, rather than receives, form. This is what he means at *GA* 729b12–14 when he says that the female qua female is passive (*pathêtikon*) while the male qua male is active (*poiêtikon*).[32]

Again, at *GA* 730a24–30 Aristotle argues that the fact that what the female contributes is 'the matter for generation' explains why she cannot generate on her own without any contribution from the male: for there must be a principle in addition to the matter 'that moves and defines it' (*archês kai tou kinêsontos kai dioriountos*). The context makes it clear that by associating menstrual blood with the matter of generation Aristotle means to pick out her contribution in its role as the passive subject of change (as something acted upon and moved).

But this does not mean Aristotle thinks the female's contribution to generation 'is empty and completely malleable with no specifications of its own' (Connell 2016, 319). As we saw in Chapter 4, the fact that something functions as matter in one context does not mean it cannot also serve as an agent of change in another.[33] This allows us to accommodate the fact that in some places Aristotle seems to assign the female a more active role in the process. For example, immediately after the passage at *GA* 730a24–30 Aristotle qualifies his claim by saying that in some cases the female can generate on her own (parthenogenesis). In these cases we can still distinguish between the menstrual blood's role as the matter for generation from its role as the efficient cause: in cases of parthenogenesis menstrual blood is acted upon and receives form qua matter (in virtue of its of moist and dry elements) but acts and imposes form qua efficient cause (in virtue of its own proper heat). Likewise, for cases of maternal inheritance: at *GA* 768a14–21 Aristotle tells us that the female's seed conveys a set of 'formative motions' (*hai kinêseis hai dêmiourgousai*) in virtue of which it is capable of transmitting family resemblances.[34] Again, we can distinguish here between the passive capacities that belong to menstrual blood qua matter from any active capacities it may have in virtue of possessing its own heat and motions.

By calling menstrual blood matter Aristotle also means to identify it with the animal offspring in a state of potentiality (*dunamei*). Matter is associated with potential being (e.g.) in *Metaphysics* VIII:

[32] See also Gelber (2010, 202).

[33] Recall in *PA* II 1 Aristotle lists the hot, cold, moist, and dry as the matter of composite bodies whereas in *Meteor.* IV the term "matter" is reserved for the two passive powers (dry/moist).

[34] Henry (2006b, 274–5).

> Since substance in the sense of subject and matter (*hôs hupokeimenê kai hôs hulê ousia*) is generally recognised and this is what exists in potentiality, it remains for us to say what the substance of sensible things is in the sense of actuality. (*Metaph.* VIII 2, 1042b9–11)

> It is clear, then, that if they follow their typical way of defining and speaking, it is not possible for them to give an account and resolve the difficulty. But if, as we say, one component [of the substance] is *matter* and the other *form*, and one is [the substance] *potentially* and the other *actually*, the issue will no longer seem puzzling. (*Metaph.* VIII 6, 1045a20–5)

In the *Generation of Animals* female matter is consistently identified with the potential substance. Since menstrual fluid is a form of the mother's own blood (which serves as the matter for her body), it already *is* her parts in potentiality or in capacity:

> It is for the sake of the better and the end that nature diverts the menstrual blood to this place [sc. the uterus] and uses it there for generation, namely, in order to generate another individual of the same sort. For [sc. being a form of her own blood] it is already potentially the same sort of thing as the body whose secretion it is. (*GA* 738a37–b4)

> The cause of this [sc. the differentiation of the parts] is the fact that the residue from the female is already potentially the same sort of animal as her offspring is naturally and that it contains its parts in potentiality, though none are present in actuality ... The matter, then, comes from the female while the male supplies the starting point of motion. (*GA* 740b18–25)

> Since the parts [sc. of the offspring] are present in the matter in potentiality, once it secures a principle from the male it develops in a connected sequence just as in the case of a self-moving automaton ... And they come to be in actuality what they previously were only in potentiality. (*GA* 741b7–8, b13–15)

Conceptualising female matter in this way provides Aristotle with a legitimate alternative to preformationism. The matter contains the parts of the offspring only as a passive capacity to be formed into them when acted upon by the appropriate agent.

5.6 The Subject of Animal Generation

According to the Subject Condition introduced in Chapter 2, all coming-to-be requires an underlying subject (*to hupokeimenon*) from which the change proceeds non-incidentally (*Ph.* 190a13–16). According to *Physics* I 7, this role is played by the seed (*sperma*) in the case of animals and plants (190b4–5). This is straightforward in the case of plants; for example, the subject from which an oak tree comes to be is the acorn. In the case of animals, however, "seed" (*sperma*) is ambiguous between male semen, female menstrual blood, and the embryo formed from these. Which of these counts as the proper subject of animal generation?

As a way into this, we may recall *GC* I 3 where Aristotle argues that substantial generation proceeds from unqualified non-being, which he conceptualises there as the sort of subject that *is* a substance in potentiality but *is not* a substance in a state of completion. According to the *GA*, this description is satisfied by unfertilised menstrual blood, which is potentially what the finished animal is in a state of completion (e.g. *GA* 738a37–b4, 740b18–25, 741b7–8, b13–15). However, there is a tension here with the *Metaphysics* IX 7 account which is supposed to tell us when we have a potential *F* (where *F* identifies the thing at the end of the change). The *GA* and *Metaphysics* agree in associating potential being with some kind of pre-existing matter from which the change proceeds. They also agree that one of the conditions for being a potential *F* is that this matter be in a state where it is capable of being changed into an *F* when acted on by the appropriate agent. But *Metaphysics* IX 7 relies on a more robust set of criteria for counting something as potentially *F* which require the matter to be in a state where it will *immediately* develop into an *F* of its own accord if nothing obstructs the process. As it happens, unfertilised menstrual blood fails this condition. For something must be added, removed (cf. *GA* 728a26–30), or changed before it is in a state where an animal will immediately come to be from it. This does not mean the two accounts are antithetical to one another. But the *Metaphysics* does provide for a somewhat different answer to the question of when we have a potential animal and thus a different answer to the question of what constitutes the proper subject of animal generation. Making the two accounts compatible requires a bit of work. It is thus worth saying something about the *Metaphysics* IX 7 account of potential being and how it relates to what Aristotle says in *GA*.

The first half of *Metaphysics* IX 7 takes up the question: 'When is a thing potentially *F* and when it is not? For it is not potentially *F* at any and every time' (1048b37–1049a1). In the case of artefacts Aristotle says that a thing counts as potentially *F* when it is in a state where nothing needs to be added, removed, or changed before it can immediately be transformed into an *F* (1049a10–11). For example, cement mix is not yet a potential foundation because we still need to add water before it is in a state where it can be used to make a foundation. Likewise, Aristotle says that earth is not yet potentially a statue because it still needs to be changed into bronze (or some other proximate material) before it is in a condition where a statue can be directly formed out of it (a17–18). The point here is that statues and foundations come to be *directly* or *immediately* out of bronze and cement, not earth and cement mix. What Aristotle wants to say in the case of naturally generated organisms – 'where the thing has the source of generation in itself' (1049a14) – is much less clear. Is an apple[35] potentially a human being? No, Aristotle says, but only when it has been converted to seed (*sperma*), though perhaps not even then (1049a1–3). Eventually, Aristotle says that seed is not the potential human until it undergoes a 'further transformation in another' (a14–15) and acquires 'another *archês*' (a16–18). For it is only then that seed is in the condition where it will straightaway develop into a human through its own principle (*dia tês hautou archês*, a15–16), if nothing prevents it.

This account is obscure to say the least. First, it is fairly clear that in *Metaphysics* IX 7 Aristotle uses "seed" to mean male semen. For when he says that this seed is not yet potentially a human because it must undergo a further change 'in another' (1049a14–15), he could only be talking about some change that the male's semen supposedly undergoes in the female. (No one would suggest that female seed, let alone the fertilised embryo, must undergo a further change in the male!) In that case *Metaphysics* IX 7 is suggesting that semen is potentially a human when it is in a state where nothing further needs to be added, removed, or changed before it will straightaway develop into an animal through its own principle. It is not difficult to see why this conflicts with Aristotle's own mature theory of animal generation. For, although semen contains the principle of generation, Aristotle is categorical in the *GA* that menstrual blood is the material out of which the new offspring comes to be after being catalysed by the heat in the father's semen, which then evaporates (737a11–16). However, it is likely that Aristotle is simply formulating his account in *Metaphysics* IX 7 using a more popular model of generation rather than his own technical theory.[36] As we shall see, Aristotle's point can be made just as easily by taking "seed" to refer to menstrual blood. But it is not *unfertilised* menstrual blood that counts as the potential human (as the *GA* theory maintains) since that is still not in a state where it will *immediately* develop into a human being of its own accord. It still needs to undergo a further change before it reaches that point: it must acquire a generative principle from the

[35] Aristotle also says earth here (1049a1) but the point is more intuitive when asking about the food an animal eats, which is the raw material that is eventually converted into *sperma* (*GA* I 18–19).

[36] Recall Aeschylus, who compared semen to a seed planted in pot (the pot being the mother).

male (*GA* 741b7–8; cf. 1049a16–17). And yet not even fertilised menstrual blood qualifies as the potential human according to the *Metaphysics* criteria. It is only when the fertilised embryo has reached the stage where it has developed a primitive heart with rudimentary nutritive capacities that it is in a state where it will *straightaway* develop into a human being *of its own accord* if nothing prevents it (*GA* II 1, 735a12–26). Then, and only then, do we have the potential human being.[37]

According to the *Metaphysics* IX 7 account, then, the proper subject of animal generation should be the fertilised embryo when it has progressed to the stage where it has developed a primitive heart endowed with basic nutritive capacities for processing blood coming from the uterus.[38]

There is a way to integrate these two accounts by drawing a distinction between different stages of generation. Consider Aristotle's statement at *Metaphysics* 1049a16–17 that the seed needs to acquire another *archê* before it is a potential human. Now Aristotle could simply be referring to the fact that the seed still needs to acquire another causal principle, namely, the efficient cause (e.g. *GA* 741b7–8). But an equally plausible translation of *heteras archês* is 'another starting point' or 'another beginning'. On this reading Aristotle is saying that the seed (in some prior state) must reach another starting point in the generative process before it is considered a potential human. Consider all of the transformations that the mother's food must undergo before it is finally in a state where it will, without further ado, develop into a new animal through its own agency. *GA* I 19–22 identifies menstrual blood as the material out of which the embryo comes to be. Menstrual blood is in turn formed out of a surplus of the mother's nutritive blood, which is itself the product of another process, namely digestion. Working backwards even further, the process of digestion ultimately begins from the food the mother consumes (e.g. an apple). Each of these different processes (digestion, spermatogenesis, embryogenesis) begins from some pre-existing matter (apple, blood, menstrual blood) which is itself the product of the prior stage:

(1) Digestion begins from the food the mother eats (Subject 1) and converts it to blood (Product 1).
(2) Spermatogenesis begins from a surplus of that blood (Subject 2) and converts it to menstrual blood (Product 2).

[37] Things might be more complicated in the case of human beings. For they possess *nous* and Aristotle says there is a "great puzzle" concerning the origins of this part of the soul, since it does not involve the development of bodily organs (*GA* II 3, 736b21–9).

[38] Gill (1989, 102–5) identifies the embryo at this stage with the proper subject of animal generation.

What *Metaphysics* IX 7 implies is that menstrual blood is still not yet a potential animal because it must undergo a further change in order to reach another starting point (*heteras archês*). That new starting point is the stage in the causal history of the matter where it has undergone a further change that transforms it into an embryo with a functioning heart:

(3) Embryogenesis begins from menstrual blood (Subject 3) and converts it into an embryo endowed with a primitive heart and rudimentary nutritive capacities (Product 3).

Following this pattern, the thing that serves as the product of this change in turn becomes the subject for the next change (cf. *GA* 739b33–7):

(4) Morphogenesis proceeds from the embryo (Subject 4) and transforms it into an organism with the same shape and form as its generating parents (Product 4).

If this is right, then unfertilised menstrual blood is the proper subject of embryogenesis, while the embryo is the proper subject from which morphogenesis proceeds (the part of the process where the embryo comes to acquire a new substantial form).[39] Still, it is reasonable that Aristotle typically speaks of menstrual blood, nevertheless, as the potential animal in the *GA*. For what makes the embryo a proper subject of animal generation is the fact that it is composed from a species-specific material that ultimately has its causal origin in the female's menstrual blood.

In this chapter I have argued that the first version of Aristotle's reproductive hylomorphism developed in *GA* I 18–22 applies to the first of these (embryogenesis). It is a way to analyse the causal contributions that male and female *sperma* make to the initial formation of the embryo, which functions as the proper subject from which the development of a new animal properly begins (morphogenesis). In Chapter 6, I turn to the second version of the model, which Aristotle uses to analyse the contributions that the mother and father make to the substance that comes to be from that change.[40]

[39] This interpretation of *heteras archês* makes good sense of Aristotle's point at *Metaphysics* 1049a21–2 that it is always the last term in the series (*to husteron*) that is more properly potentially *F*.

[40] I shall have more to say about (4) morphogenesis in Chapters 7 and 8.

CHAPTER 6

Biological Generation: Part Two

6.1 Body and Soul

In the previous chapter I argued that in the context of *GA* I Aristotle uses the hylomorphic model primarily as a way to analyse the causal contributions of male and female seed to the formation of the primitive embryo (embryogenesis). In this context semen is said to give form to the menstrual blood only in the purely mechanical sense of imposing a determinate limit or boundary on her indeterminate fluid. In *GA* II 4–5 Aristotle uses the hylomorphic analysis as a way to identify the parents' reproductive contributions to the substance that comes to be from the embryo (the completed offspring). He presents his account in two stages. In *GA* II 4 he argues that the father provides the offspring's form in the sense of its soul while the mother provides the body that underlies the soul as its subject. He then refines this account in *GA* II 5 by identifying the father's *exclusive* contribution with *sensory* soul. At the same time, Aristotle allows that some of the offspring's nutritive capacities are acquired directly from the mother without any input from the male. Thus, if form (*eidos*) is taken in the narrow sense to refer to essence, then there is a sense in which the father *alone* provides the animal's form insofar as the capacity for sensation is that feature that makes a living thing an animal. But if biological form is taken in the broad sense to pick out soul-capacities in general (*DA* II 1), then both males and females can equally be said to make a formal contribution to the substance that comes to be from them. This will take some fleshing out.[1]

[1] At this point it is necessary to head off a potential objection. It is true that Aristotle says, at the end of *GA* II 3, that females cannot generate on their own on the grounds that menstrual blood lacks 'the principle of the soul' (737a25–33). But this is compatible with my reading of *GA* II 5. For Aristotle is explicit in that later chapter that what he means is that her generative residue lacks the source of the sensory part of the soul. This is why she cannot generate *an animal* on her own, since the capacity for sensation is what makes something an animal. See Mayhew (2004, 43).

In *GA* II 4 Aristotle appeals to the hylomorphic model to explain the fact that not all males discharge semen even though all females produce some kind of spermatic material:

> The reason why (*aition*) a generative residue is not present in all males while it is in all females is that the animal is an ensouled body (*to zôon sôma empsuchon*). The female always provides the matter, while the male is the agent that fashions it. For these are the powers of each (as we have said) and what it is to be female and male. So that while the female necessarily provides body and bulk, it is not necessary for the male to do so. For it is not necessary either for the tools or the producer to be inherent in (*enhuparchein*) the things that come to be. Instead the body comes from the female, while the soul comes from the male; for the soul is the substantial being of a certain sort of body. (738b18–27)

Just before this passage Aristotle had argued that menstrual fluid comes to be in females owing to so-called material necessity (738a33–b4). All females have a principle of natural heat that is colder and weaker than the male's, which in this context refers to the inability to effect concoction (hotter/colder and stronger/weaker being measures of the power of an animal's natural heat to do work). Because of this she is unable to concoct the excess blood that collects in the blood vessels around the uterus, and so the vessels 'become overfilled with blood'.[2] And since the vessels are extremely narrow, the excess of inconcocted blood eventually overflows into the uterus. The upshot of this is that, even before the male makes his contribution, there is already in the place where generation takes place some matter available for use, which is 'potentially the same sort as the body of the animal whose secretion it is' (738a37–b4).[3]

This sets up the main puzzle of *GA* II 4: 'Why, although this sort of generative residue is present in all females, is it not present in all males?' (738b9–27). Aristotle goes on to argue that the reason for this is that an animal is a composite of body (matter) and soul (form). And one parent must provide the body while the other provides the soul. On the one hand, Aristotle has just shown that all females necessarily supply a residue that

[2] According to *Meteor.* IV 2–3 concoction makes things smaller, denser, and more compact. This is why the vessels in question become overfilled with blood. Unlike the male, the female's principle of natural heat isn't strong enough to concoct the blood and reduce it to a smaller quantity.

[3] This teleological account is not part of the explanation for why that generative residue is present in the first place. Its presence in the uterus is a necessary consequence of the female's inability to concoct her blood together with the structural limitations of the blood vessels surrounding the uterus. This necessary residue is co-opted for the sake of reproduction, but it did not come to be for that reason.

can be used as the matter for generation. On the other hand, he argues that it is not necessary for the agent (the male) to supply anything material to the generated product (738b24–5, cf. 729b9–11). For example, the doctor does not heal the patient by grafting some of his own healthy flesh onto the patient's body. Likewise, when the builder builds a house no part of the builder's body serves as an ingredient in the house being built (cf. 730b8–32). The argument thus runs as follows:

(1) The living animal is a composite of body and soul. (Hylomorphism)
(2) In generation one part of this composite comes from the female while the other comes from the male. (Reproductive Division of Labour)
(3) Of necessity there is some material residue in the place of generation that comes from the female ('the female necessarily provides body and bulk'), which is potentially the same sort of body as her own.
(4) It is not necessary for the male (as agent) to provide anything material to generation.
(5) Therefore, the body of the offspring comes from the female while the male provides soul, which is the form and substance of that sort of body. (738b25–6)

In this context, then, the form that comes from the father is the offspring's soul while the matter that comes from the female is associated with the body that underlies that soul as its proximate subject.[4]

Next Aristotle appeals to the case of hybrids in order to confirm his view that females provide the matter (or body) for the offspring, which reinforces the conclusion of the previous argument that males provide its form (or soul) (*GA* 738b27–739a2). He notes that when two closely related species reproduce, at first the hybrid takes after both parents, but after several generations the offspring comes to resemble the mother's species 'with respect to its outward appearance' (*kata to thêlu tên morphên*). What Aristotle has in mind here are the effects that the animal's diet has on its

[4] This way of reading Aristotle's reproductive hylomorphism has been challenged recently by Kosman (2010). Instead, Kosman argues that Aristotle's account is meant to identify the parent's contributions to the form and mater *of the process of generation itself.* But the *GA* II 4 passage clearly means to apply the form–matter analysis to the contributions that each parent makes, not to some aspect of the change, but to 'the ensouled body' (*to zôon sôma empsuchon*: 738b19–20) that results from that change. And the form supplied by the father is explicitly identified with its substantial being (*ousia*). There is no mention here (or anywhere in the *GA*) of the male giving form to its coming-to-be (*genesis*). However true it may be that the specific pattern of changes development follows owes its "form" to the generating parent, Aristotle never associates this with the *eidos* identified by the hylomorphic model of reproduction.

colour (e.g. *GA* V 6, 786a34–b5) and various other bodily features that are dependent on its material constitution.

If this is right, then the appeal to hybrids does not show that Aristotle thinks the female also determines what type of soul the offspring has (Connell 2016, 155). Nor does it show that he was open to the possibility of an evolution of species (pace Balme 1972, 97). Aristotle is clear that the hybrid eventually *reverts* to the species of the female. For example, if a female wolf and a male coyote cross, the first few generations will produce coywolves (the hybrids). But eventually the lineage will return back to wolves. So no new species are permanently created through this process. The possibility of an evolution of species is not on Aristotle's radar here. What he is focused on is the fact that the matter an animal consumes can affect certain features of its appearance (e.g. *GA* 786a34–b5). This is why it is significant that hybrids eventually revert to the mother's species with respect to its bodily appearances. For it helps show that the matter comes from the female rather than the male. Whether or not this change constitutes a permanent evolution of species is not relevant to the argument taken in context.

But the *GA* II 4 account does not give us the whole picture. When we get to *GA* II 5 we learn that what males alone contribute to the generation of an animal is not the whole soul but rather those sensory capacities that constitute the form of the animal in the strict sense: 'Hence, in such animals [sc. that are divided into males and females], the male always brings generation to completion (*epitelei tên genesin*); for he implants (*empoiei*) the capacity of the sensory soul, either directly or through the semen' (*GA* II 5, 741b6–7). How Aristotle arrives at this conclusion will be important for assessing the full impact of *GA* II 5 below. For now, we can simply note that, in its most developed form, the second version of Aristotle's reproductive hylomorphism identifies the capacity for sensation as the defining contribution of the male animal. And since it is the possession of sensation that makes the ensouled body *an animal*, this amounts to the claim that males alone provide the animal's form in the sense of its essence. It is in this sense that the male provides the formal (and final) cause of the animal. Since animals are defined by the possession of sensation (e.g. *De juv.* IV 469b3–5), it follows that no living thing could *be* an animal without this part of the soul. This is ultimately what explains why females cannot generate on their own; for what they would generate would not be *an animal* by definition but something that 'lives the life of a plant' (736a34, 736b12–13). The argument of *GA* II 5 is critical for my interpretation, so it is necessary to go through it with some care.[5]

6.2 What Is the Point of Males?

Aristotle opens *GA* II 5 by raising the following puzzle about the role of males in generation:

> Now in plants the female is not separate from the male. In those animals in which the female is separate, there is a need for the male in addition to the female. And yet one might raise a puzzle (*aporia*) about the reason for this. Since the female has the same soul as the male, and since the matter is the residue supplied by the female [sc. which comes to be of necessity and is already potentially the animal], why is there a need for the male in addition? Why doesn't the female simply generate the offspring from herself by herself? (741a3–9)

This question is quite unexpected if Aristotle meant to identify the female's contribution exclusively with matter. For on his account generation requires matter *and* form. So, the fact that he even raises this puzzle here should already alert us to the fact that the hylomorphic model presented in *GA* II 4 is not the complete story.

As mentioned, Aristotle's answer to this *aporia* is that males are necessary for generation because they provide the sensory capacity of soul, which is the property that makes the offspring an animal:

> Why doesn't the female generate the offspring from herself by herself? The reason is that there is a difference between an animal and a plant. The animal possesses sense-perception. It is impossible for a face or a hand or flesh or any other part <of the animal body> to exist without having sensory soul in it, either in activity or merely in capacity, and either in some particular sense or absolutely. For <without sensitivity> it will be like a dead body or a part of a dead body. If, then, the male is that which is productive of this kind of soul, then wherever male and female are separated, it is impossible for the female to generate an animal from herself by herself. For this, as we have said, is what it is to be male. (*GA* II 5, 741a8–16)

[5] By saying the male alone provides sensory soul I do not mean he does so entirely by himself without any input from the female (Connell 2016, 177 n. 44). My reading assumes there is a suitable matter from the female present for the male's contribution to act upon with all the right sorts of passive capacities. The point is rather that Aristotle thinks there is a sense in which it is true to say of the male *and not the female* that he brings generation to completion insofar as he alone implants the sensory soul (741b6–7).

I take Aristotle's point in this passage analytically. If an animal is (= def.) an organism that possesses the capacity for sensation, and if the male alone is responsible for providing that capacity, then it stands to reason that if the female could generate an offspring by herself what she would generate would not *be* an animal *by definition*. It would be like a dead body, which is an animal "in name only". It is important that Aristotle does not say her offspring would literally be a dead body; he isn't saying she would always have a still birth. As we shall see, Aristotle does think that females can generate by themselves something that has life. But it would be an animal in name only because it would lack that feature that makes something an actual animal.

Aristotle's argument for claiming that males alone provide the sensory capacities of the offspring's soul turns on the principle that nature does nothing in vain (741b4–5).[6] If the female could generate an animal all by herself, then the male would exist in vain. But nature does nothing in vain. Therefore, the male must be making some necessary contribution to the generation of an animal that is not provided by the female. In order to identify the male's exclusive contribution Aristotle turns to cases of parthenogenesis. Since the parthenote ("wind-egg") is generated by the female without any reproductive input from the male, these cases can help to isolate what it is that males bring to the generation of an animal. In those cases, Aristotle tells us that the female's own contribution contains enough vital heat to form the primitive embryo and catalyse its development (cf. *GA* I 21, 730a30–2). However, observation shows that the parthenote can only develop up to a point before it is spontaneously aborted.[7] Aristotle thinks that this is the point where normal embryos acquire sensitivity, which suggests that the sensory capacities of the offspring's soul must be coming from the male.

But parthenogenesis also revealed something else to Aristotle. It provided him with strong empirical evidence that the offspring must also be getting some of its soul-capacities from the female. For the parthenote grows and decays. And anything capable of engaging in those activities must be alive and possess soul (cf. *DA* II 1, 412a14–15):

[6] For a discussion of this principle see Henry (2013) and references therein.

[7] Aristotle's account is more or less correct, but his experience with parthenogenesis seems to have been limited to birds. Complete development of a parthenote individual is relatively rare in vertebrates, including birds. Schut et al. (2008, 197) report that in those species of bird where parthenogenesis has been observed 'development was disorganised and almost always abortive'. Yet, while rare, complete development has been empirically verified (e.g. Kono et al. 2004), which effectively undermines Aristotle's central argument for the *GA* II 5 version of the hylomorphic model.

> What sort of soul, then, do these [female induced] eggs possess? It must be the lowest level, namely, the nutritive soul. For this is present in plants and animals alike. Why, then, does this soul [sc. the one possessed by the parthenote] fail to bring the parts to completion and make an animal? It is because the parts of an animal must have the capacity for sensation; for the parts of animals are not like those of plants. This is why the male must have a share in generation in those species where the male is separated. (*GA* II 5, 741a24–8)

Aristotle repeats the point later in Book III:

> Wind-eggs acquire completion as far as is possible for them. It is impossible for them to become complete to the point of being an animal, for that requires possessing sensation. But, as has often been said, the nutritive power of the soul is possessed by both females and males as well as all living things. Hence, the embryo itself, *considered as a plant*, is complete [sc. because it has nutritive soul], but, *considered as an animal*, it is incomplete [sc. because it lacks sensory soul]. (*GA* III 7, 757b14–19)

The point of Aristotle's appeal to parthenogenesis is clear. Unfertilised embryos develop up to a certain point, which they do by means of basic nutritive capacities acquired directly from the female. But these embryos do not develop to the point of being an animal, which shows that those capacities are not sufficient by themselves to complete development. They can only produce something that lives the life of a plant (something that can only engage in nutritive activities). Aristotle takes this as empirical proof that males alone contribute the sensory capacities of the soul.[8]

According to *GA* II 5, then, there is a sense in which both parents make a contribution to the offspring's form insofar as they both contribute capacities of soul. This is only incompatible with the *GA* II 4 argument if we take Aristotle to be saying there that the father provides all capacities of soul. But the argument need not be read in that way. Aristotle may simply mean that the kind of thing the male provides to the offspring are soul-capacities as opposed to anything bodily. That does not commit him to the claim that the male provides all of its soul-capacities. Nor does the *GA* II 4 argument commit Aristotle to the claim that the female only provides the body of the offspring; rather, he says (i) it is necessary for her to do so and (ii) she alone does so. Thus, on my interpretation, *GA* II 4 does not say that males *alone* provide soul-capacities. Aristotle's point in 738b18–27 is that, of

[8] My interpretation of *GA* II 5 only commits Aristotle to the minimal claim that *some* of the animal's nutritive capacities are supplied directly by the mother, though it is consistent with the stronger reading according to which *all* nutritive capacities trace their causal history to the female's formal contribution.

the two components of the ensouled body, the male's contribution is associated with the soul *and not* the body. *GA* II 5 can then be read as further specifying more precisely which capacities of soul are the defining contribution of the male. It says that what males alone provide are the sensory capacities of the offspring's soul, which is the substantial being (*ousia*) of the animal body in the strict sense of its essence. But *GA* II 5 also argues, on good empirical grounds, that (some of) the nutritive capacities of its soul must be coming from the female.[9]

Not all commentators agree on the significance of parthenogenesis in *GA* II 5. For example, Jessica Gelber argues that Aristotle's remarks there do not actually support the view I attribute to him:

> While Aristotle does say that wind-eggs [parthenotes] have soul, he adds that it is clear that they have soul only potentially (741a23). Wind-eggs do not actually have nutritive soul. There are living organisms that have nutritive soul in actuality but lack sentient soul, namely plants. But Aristotle does not think that the mother makes a plant that can then become an animal once sentient soul is added. (Gelber 2010, 200)

But this is not quite right. My reading does not entail that wind-eggs are literal plants. What Aristotle says is that they 'live the life of a plant' (*phutou bion zên*: e.g. 736a34, 736b12–13; cf. 753b25–8, 779a1–2). At the end of the *GA* III passage (quoted above) Aristotle expands on this by saying that *considered as a plant* the parthenote is complete because it has nutritive soul but *considered as an animal* it is incomplete because it lacks sensory soul. It is clear what he means by this. The parthenote is like a plant insofar as it is capable of processing nutriment and growing bigger, activities that are characteristics of plants. For Aristotle, this implies that they have some measure of nutritive soul in them, which could only have come from the mother since the parthenote has no contribution from the father. Whatever Aristotle means by saying that parthenotes have nutritive soul 'in capacity' (*dunamei*), it must mean they possess soul to some degree since they grow and decay. And Aristotle is clear that the soul is the source (*archê*) of these life activities (*DA* II 2).

Gelber's other reason for denying that parthenotes actually possess nutritive soul is that they cannot become animals once sentient soul is added to them. This is not right either. While the parthenote lives the life of a plant, Aristotle is clear that they will develop into a complete animal if they are later fertilised by a male as long as it happens within a certain

[9] Peck (1942, xii–xiii) defends a similar interpretation.

period (*GA* I 21, 730a4–17; III 1, 750b26–34; III 7, 757b1–30; *HA* VI 2, 560a9–12). My reading makes sense of this. At first the parthenote's soul is limited to those nutritive capacities it receives directly from the mother. This is why it can only develop up to a point before it is spontaneously aborted. If it gets fertilised by a male before that point, it will continue to develop into a mature animal since it will have acquired those capacities necessary to complete its development: 'For in such animals the male always brings generation to completion. For he implants the capacity of the sensory soul, either directly or through the semen' (*GA* II 5, 741b6–7).

6.3 The Unity of the Soul

Someone might also raise an objection to my interpretation of Aristotle's reproductive hylomorphism on the grounds that it is incompatible with the unity of the soul, which is an important part of Aristotle's account in the *De anima*. For example, Connell (2016, 173) says that 'it presupposes that nutritive and sentient souls [soul-parts?] are in some quite robust way separable'. According to the traditional reading, although Aristotle talks about various "parts" of the animal soul – the nutritive part, the sensory part, and (in some animals) the rational part – these do not constitute *separate* parts of composite soul. Instead they refer to different *modes* or *aspects* of a single unified capacity for life. To talk of the nutritive faculty of the soul is to talk about this single capacity operating in its nutritive mode, while talk of the sensitive faculty characterises it in its perceptual mode. On this reading there are "parts" of the soul only in the weak sense that the left and right sides of a sphere can be called "parts" (aspects) of the sphere. Joachim expresses this sort of view when he writes:

> The animal-soul, whose distinctive character is its sentience, also functions as assimilative and reproductive; but it is not the nutritive faculty of the plant-soul *plus* the sentient faculty. The soul is always – as the form of its body – individual, a unity: and though we distinguish different functions as modes in which its unity is manifested, they are not separate or separable. (Joachim 1951, 39)

If the capacities for sensation and nutrition are not separate parts of the animal soul but merely different modes of its existence, then (the objection goes) to say the father provides the sensory part entails that he provides the nutritive part. For there is only one indivisible soul to give, which comes from the father.

A similar objection could be raised on the basis of what Aristotle says about the relation between parts of the soul in *DA* II 3. There he presents a model of the soul in which the lower parts are contained potentially in the higher parts in the way that the triangle is contained in the quadrilateral (414b28–415a13). The quadrilateral is not composed of the triangle as a separate part; rather, the triangle is only a potential part of the quadrilateral that contains it. Once again one might take this to show that the animal soul just is the sensory soul, which contains the lower-level nutritive soul only in potentiality.[10] One way to cash this out is to take the nutritive soul and the sensory soul as the same soul in different stages of completion. In support of this one could point to *GA* II 3 where Aristotle argues that at first embryos possess a primitive nutritive soul and then later develop those sensory capacities that makes them an animal (736a35–b5). On this reading, the animal soul is the perfection of the embryo's more primitive nutritive soul. This seems to be what Joachim has in mind when he goes on to characterise the nutritive faculty as 'a subordinate power, taken up into and fused with the higher power (the sensitive faculty, *to aisthêtikon*) which characterises and constitutes their more developed soul'. And so again there is only one soul to give, which must come from the male.

This second objection can be dealt with swiftly by noting that Aristotle thinks matter and form also constitute a unity that depends on the same potentiality–actuality relation from *DA* II 3 (*Metaph.* VIII 6). And he is clear that these two parts of the composite have different reproductive origins, the matter coming from the female and the form coming from the male. So the claim that the nutritive and sensory parts form a unity that depends on standing in a relation of potentiality to actuality does not rule out their having different reproductive origins.

We can respond to the first objection by noting that the traditional interpretation (which treats nutrition and sensation as different modes or aspects of a single unified capacity for animal life) is not the only way that scholars have understood Aristotle's talk of soul-parts. In *DA* II 2, for example, Aristotle considers two ways that the different capacities of soul might count as separable parts: they might be separable only in account or also separable in place.[11] *A* and *B* are said to be *different* in account if the account of *A* is simply different from the account of *B*. *A* is *separable* in account from *B* if the definition of *A* makes no reference to *B* (though not necessarily vice versa). According to Corcilius and Gregoric (2010), Aristotle treats separability in account as the criterion for soul-parthood.

[10] See Shields (2016, 197–8), Johansen (2012, 67–72).

[11] For a discussion of how these types of separability relate to the parts of soul see Corcilius and Gregoric (2010).

For example, while imagination differs in account from sensation (they have different definitions) it is not *separable in account* from it, since its definition includes reference to sensation as its cause (*DA* 429a1–2). By contrast the definition of sensation makes no reference to imagination and so is separable in account from it. Thus, imagination cannot be treated as a separate *part* of the soul; rather, it is a capacity of the sensory part (whose account makes no reference to any other capacity). On this reading, the capacities for nutrition and sensation constitute proper parts of the animal soul, since each of these is separable in account both from one another and from those other capacities whose accounts ultimately make reference to one of them.

Whiting (2002) argues for an even stronger notion of separable parts according to which the parts of the soul are separable, not only in account, but also in place (or magnitude) insofar as each part is embodied by a different system of physical organs and is thus located in different portions of matter. For example, the nutritive part is embodied in that physiological system that underwrites the activities of nutrition and generation, while the sensory part is embodied in a different system of organs. According to Whiting (2002, 152) this makes the parts of the animal soul spatially separate from one another insofar as the physiological systems that embody those parts occupy different places in the animal's body.

Even if this is right, one might still worry that my reading implies that the parts of the soul are capable of *existing* in separation from one another – the nutritive part in the menstrual blood, the sensory part in the semen – which Aristotle denies.[12] While the nutritive capacities of soul might be capable of existing in separation from the sensory capacities, as they do in plants, the sensory capacities cannot exist apart from the nutritive ones. And so, if the father contributes the sensory part (the objection goes), then he must also be contributing the nutritive part. For the sensory part of the animal soul is not capable of existing in the semen in separation from the nutritive part.[13]

But this misunderstands how Aristotle thinks semen transmits soul. In the *De anima* a soul is defined as the first-actuality (developed capacity) of

[12] I take it this is what Connell (2016, 173) means by being 'in some quite robust way separable'. The following is meant to answer that objection as well as her second objection that this interpretation implies 'that a portion of soul can be handed over or supplied by the male parent'.

[13] A stronger objection would hold that, even if the case of plants shows that nutritive capacities can exist without sensory capacities in general, the nutritive capacities of animals differ in kind from those that constitute the plant soul. And so the case of plants is not enough to show that Aristotle thinks the nutritive part *of the animal soul* is capable of existing in separation from the sensory part. My reply covers both the stronger and weaker versions of the objection.

a natural body equipped with organs specially adapted for executing the functions associated with that kind of soul (an *organikon* body).[14] This is why Aristotle says in *GA* II 3 that those capacities of soul whose activity is bodily cannot belong to anything except that of which it is the body, e.g. the capacity to walk cannot belong to anything except feet (736b22–4). But semen is not the sort of body (nor does it contain the sort of body) that could possess a sensory soul in that sense. Instead, what the semen possesses is a certain power (*dunamis*) for generating sensory organs whose own developed capacities constitute the sensory part of the future offspring's soul.[15] This account of how the parts of the soul are transmitted in reproduction is grounded in Aristotle's theory of heredity in *GA* IV 3. I discuss this in Chapter 7. According to that theory, semen has this power in virtue of the "motions" it contains. Likewise, Aristotle argues that the menstrual blood is endowed with a set of spermatic motions in virtue of which it transmits the nutritive part of the soul (also as a power for generating the corresponding system of organs). If this is right, then the father's semen is responsible for endowing the offspring with sensory soul, not by implanting some kind of latent soul directly into the female's matter, but by endowing the matter with a set of active powers which (in conjunction with its own passive powers) cause it to develop the sorts of organs whose developed capacities constitute the sensory part of the offspring's soul.[16] This avoids the unity objection, since it does not commit Aristotle to the view that the sensory part of the soul *exist* in the semen in separation from the nutritive part that comes from the mother. For the parts of the soul do not actually *exist in* the parents' seeds in any way that would threaten their unity.[17]

[14] *DA* II 1, 412b1–6. Aristotle compares the soul to the *capacity* of an axe to chop and the *power* of the eye to see (412b12–24, 413a1).

[15] Kosman (2010, 149). [16] Connell (2016, 172–7).

[17] *GA* II 3 might be taken as counterevidence to my claim that semen only transmits sensory soul. There he tells us both '*ta spermata* and *ta kuêmata tôn zôôn*' possess nutritive soul (e.g. 736a31–6), which Peck (followed by Connell 2016, 173–4) translates as 'the semens and the fetations of animals'. In that case Aristotle would seem to be saying that semen possesses the nutritive part of the soul in addition to the sensory part. However, it is more likely that *ta spermata* refers to plant seeds while *ta kuêmata tôn zôôn* means the embryos of animals. What Aristotle is saying, then, is that both plant seeds and animal embryos (which are equivalent entities) possess nutritive soul. He goes on to say that these entities only possess nutritive soul 'in capacity' (*dunamei*); they do not possess it 'in activity' (*energeia*) until they begin to process nutriment. (I take these to refer to the first- and second-actuality, respectively, i.e. to *the capacity* to process nutriment and the *active exercise* of that capacity.) Compare the argument at the end of *GA* II 1 (discussed below). More importantly, *GA* II 3 also insists that at this early stage the animal embryo does not yet possess sensory soul *even in capacity* but acquires it later on 'as it develops' (*proionta*, 736b1). The rest of the argument makes it clear that Aristotle means as it develops the appropriate sensory organs (736b13–29). This is why he

In this chapter I have developed a reading of Aristotle's reproductive hylomorphism in which both parents make a contribution to the offspring's form insofar as they both contribute parts of its soul. What males alone provide is the sensory part of that soul, which is the form of the animal in the strict sense of its essence. It should be clear by now that this interpretation does not end up being dependent on any particular view concerning the unity of the soul in the mature animal as some have suggested.[18] For my interpretation is only a claim about the reproductive origins of the different parts of the soul and how those parts are transmitted during reproduction. It does not commit Aristotle to any particular view about how those parts are ultimately related to one another in the mature animal or how they come to be integrated into a unified soul during the process of development. Finally, I have argued that there is no worry here about the parts of the animal soul *existing* in separation during their transmission. For the different parts of the offspring's soul do not 'exist in' the parents' seeds in any robust sense that would threaten its eventual unity. Instead, on my reading, each parent's soul-contribution is much more indirect in that what its seed transmits are "motions" whose function is to endow the embryo with a set of developmental capacities (or *dunameis*) for forming a system of organs whose own functions eventually constitute the offspring's mature soul.

says that in the beginning animal embryos 'live the life of a plant': they have the capacity for nutrition but do not yet have the capacity for sensation (they do not have it even as a first-actuality). Given all of this, it is difficult to see how Aristotle would be willing to accept that male semen possesses the sensory part of the soul except as a power to generate an animal with functioning sensory organs.

18 Again, see Connell (2016, 173) who objects that on my reading the parts of the soul end up being 'in some quite robust way separable'.

CHAPTER 7

The Efficient Cause of Animal Generation

7.1 A Puzzle about the Efficient Cause

As we have seen, Aristotle opens the *Generation of Animals* by announcing that he has already dealt with three of the four causes of animals and their parts (formal, final, and material) and that it remains to discuss the efficient cause. For 'to inquire into this and to inquire into the manner of generation for each thing is, in a way, the same thing' (*GA* I 1, 715a1–18). But here we immediately encounter a puzzle. On the one hand, the *GA* routinely identifies the male principle as the primary efficient cause of the animal and its parts, which is housed in another individual of the same species, namely the father. This would suggest that generation has an *external* efficient cause (external to the thing that comes to be from it). This view is consistent with *Metaphysics* VII 7 where Aristotle sets out his hylomorphic model of natural generation:

> Of things that come into being some come into being by nature, some by craft, while others come to be spontaneously. But in every case the things that come into being come to be by the agency of something [mover], out of something [matter], and come to be something [form] . . . The generation of natural things includes all those whose generation proceeds from a nature . . . In general, that from which they come to be is a nature as well as that in accordance with which they come to be; for what comes into being (e.g. a plant or an animal) has a nature. But so too is that by which they come to be: this is what is called the formal nature (*hê kata to eidos legomenê phusis*), which is specifically the same <as the thing generated> though it exists in another individual; for a human generates a human. (*Metaphysics* VII 7, 1032a12–25)

This passage is clear that the efficient cause of animal generation (that 'by which' it comes to be) is the formal nature of the generating parent. This view is reflected in those places in the *GA* where Aristotle identifies the father as the agent that fashions (*to dêmiourgoun*) the offspring using his semen as a tool (e.g. 730b8–32, 738b20–5).[1] On the other hand, *Physics* II 1 distinguishes natural things from the products of craft on the grounds that the former have *within themselves* a principle of motion and rest whereas the latter 'have no inborn impulse to change (*hormê echei metabolês emphuton*)' (*Ph.* 192b8–10). Aristotle defines nature, accordingly, as 'a principle and cause of being moved and being at rest *in* that to which it belongs primarily, in virtue of itself, and non-incidentally' (192b20–3). Most scholars take Aristotle to be defining nature here as an internal efficient cause. And so we should expect animal generation, as a paradigm natural change, to be efficiently caused by the nature in the embryo itself. The aim of this chapter is to reconcile these two perspectives on the efficient cause.

7.2 The Definition of Nature in *Physics* II 1

This puzzle about the efficient cause of animal generation is connected with a recent debate in the literature concerning Aristotle's definition of nature in *Physics* II 1. As mentioned, most scholars take Aristotle to define nature there as an internal efficient cause. On this reading, living things differ from the products of craft by having within themselves the efficient cause of their own production, while artefacts are brought into being by a motive principle located in something else (or in the thing itself qua other).[2] Some recent scholars have called this reading into question on the grounds that many of the changes that Aristotle considers natural have external efficient causes (e.g. the motions of the elements to their natural places). And we have seen that there are good grounds for thinking this is true even in cases of animal generation (e.g. *Metaph.* 1032a24–5). Defenders of the traditional reading might point to *Physics* 193b13–18, where the nature that gets identified with form is the nature *in* the generated thing itself. However, Stavrianeas (2015, 59) argues that this should not lead us to conclude that the developing embryo thereby possesses its own internal efficient cause. The efficient cause of the change is still assumed to be the form of the substance that generated it (as it is at 193b6–12). The nature that is in focus at 193b13–18 (the nature in the embryo) is instead operating as an internal final cause.

Kelsey (2003, 71) takes this even further and argues that it is impossible for animal generation to have an internal efficient cause since nothing can

[1] See Cooper (1988, 30), Lewis (1994, 249), Furth (1988, 117–19).
[2] For a classic statement of the traditional view in the context of animal generation see Code (1987).

generate itself, as Aristotle himself points out in the closing argument of *GA* II 1: 'The cause that is responsible (*aition*) for generation is not any part of <the offspring itself> but the primary external mover (*to prôton kinêsan exôthen*); for nothing generates itself, though once it has been generated it makes itself grow' (735a12–14; cf. 735a3–4). I will return to this passage below. As we shall see, a lot turns on how we interpret *genesis* and *auxêsis* ('generation' and 'growth') there. For now, we can note that if Kelsey is right, then this passage would be decisive in showing that animal generation must have an external efficient cause. Nothing can be the efficient cause of its own coming into existence, since it does not exist until after the process is completed.[3] The other goal of this chapter is to see what light the *Generation of Animals* can shed on this debate surrounding the definition of nature.[4] I want to make the case that there is good evidence in the biological works to support the traditional reading of *Physics* II 1 that the definition of nature is intended to be taken there in the sense of an internal efficient cause.

7.3 Natures in the Biological Works

I want to start by setting aside a passage from *GA* II 4 that has become the focus of some recent scholarship:

> Just as the things that come to be by craftsmanship are formed by tools – or it is more accurate to say by the movement of these; for this is the activity of the craft, and the craft is the form of the product in another – so too with the power of the nutritive soul: just as later on in the animals and plants themselves, this power effects growth from food using things like heat and cold as tools (for the process occurs in these [sc. in the heatings and coolings], and each thing comes to be according to a certain formula), so too this power constructs the naturally generated organism at the beginning.[5] For, since the matter with which the organism grows and the matter out of which

[3] Aristotle appears to make the same point in the *Movement of Animals*. There 'growth' seems to refer to the whole process of generation (*MA* 700a35–b3).

[4] According to Kelsey (2015), *Physics* II 1 means to define nature exclusively in terms of an internal *final* cause that initiates change as an end or goal while itself remaining unchanged. On Kelsey's reading, natural things ultimately differ from artefacts in being the proper subjects of the good for whose sake the change takes place (see Kelsey 2003, 78). For example, the good for whose sake the parts of animals come to be is the animal itself, whereas the good for whose sake the house is built is not the house but something else, namely, humans. For a challenge to the view that artefacts lack intrinsic ends see Witt (2015).

[5] See *GA* II 6, 743a36–b18. We have already encountered this idea of nature 'using' heat and cold as 'tools' in connection with *GC* II 9 (Chapter 4). It is also an important part of what I shall call the "architectonic" model of generation discussed in Chapter 8.

> it is first constructed are the same, so too the active power (*hê poiousa dunamis*) that operates at the beginning is the same [sc. as the power that maintains the body later], though it is greater. If, therefore, this is the nutritive soul, then it is also the power that generates. And this is the nature of each (*hê phusis hê hekastou*) that is present in all plants and animals alike, while the other parts of the soul belong to some living things but not others. (*GA* II 4, 740b25–741a3)

While this passage is tantalising in its details, it is ambiguous with respect to the puzzle under consideration. The main thesis of this passage is that the same power of soul is responsible both for constructing the parts of the organism at the very beginning and for growing those parts later on (see also *DA* II 416a19: 'the same capacity of soul is both nutritive and generative'). Aristotle's argument for this appears to depend on the principle from *DA* II 4 that activities are individuated by their objects and capacities by their activities (415a14–22). Blood, which is nourishment in its final form, is the matter of nutrition; female seed, which is a form of blood, is the matter of generation. Since the material objects of these two activities are the same, it follows that their corresponding powers or capacities are the same. At the very end of the passage Aristotle identifies this nutritive-cum-generative capacity of soul with the nature of each organism.

But this argument is ambiguous between two readings. On the one hand, Aristotle could be saying that *the offspring's own* nature is responsible for generating its parts at the beginning and for growing them later on. This is the reading I favour. But it is not the only way to take the passage. Aristotle also says in *DA* II 4 that the same capacity of soul is both nutritive and generative (416a19). And most scholars take that claim to indicate that the nutritive soul is both a power to maintain one's own form (nutrition) *and* a power to reproduce that form in another individual (reproduction) (416b9–20).[6] Aristotle could be making this same point in our *GA* II 4 passage. On this reading when Aristotle says the same power that effects growth from food using heating and cooling as tools is also the power that constructs the naturally generated organism at the beginning, he has in mind the power of *the father's* nutritive soul.[7] That would certainly support

[6] Menn (2002, 120–7).

[7] Lefebvre (2019) takes this reading to be implied by Aristotle's Greek. Both Lefebvre and Lennox (2018) argue that Aristotle uses the word *gennêsis*/*gennêsthai* (double nu) consistently for reproduction (the act of generating another individual like oneself). Since the *GA* II 4 passage refers to the generative power in question as *hê gennôsa* (740b37), it would follow that Aristotle has in mind the reproductive function of the father's nutritive soul (cf. *DA* 416b9–20). But Aristotle does not use the double nu consistently to refer to reproduction as opposed to generation. For example, *GA* 726b19–21 switches from *geneseôs* to *gennêtikên* without any apparent change in meaning. Again, *GA*

the Kelsey–Stavrianeas view that natural generation is efficiently caused by a principle of change in something other than the thing that undergoes the change (as he says in *Metaphysics* VII 7).

A more promising place to look is back in *GA* II 1 where Aristotle himself identifies our puzzle about the efficient cause as one of the central *aporiai* of biological generation:

> There is a great puzzle (*aporia pleiôn*) here about how on earth plants and animals come into to being from seed (*ek tou spermatos*). For it is necessary that what comes into being comes to be out of something [matter], by the agency of something [mover], and comes to be something [form] . . . The question being investigated now is not out of what but by what agency do the parts come into being. (*GA* 733b23–32)

The full resolution of this 'great puzzle' does not emerge until the very end of *GA* II 1, the core of which is contained in the passage at *GA* 735a12–26 (translated below). In this section I want to put that passage into its broader context by examining the first part of Aristotle's argument at *GA* 733b32–734b19.

The question Aristotle raises here is: what is the agency that makes the parts of the offspring? He begins his argument by considering two (apparently mutually exclusive) options: either something external to the embryo makes the parts or something existing in the embryo itself. He immediately rules out the first option on the grounds that all efficient causal interactions require contact between agent and patient, and the external agent (the father) is not in contact with the parts of the offspring as they develop. Therefore, the offspring must be fashioned by something that is present in the embryo (*kuêma*) from the start, which is either a part of it or something separate from it (734a5–6). Aristotle then launches into an argument to show that what makes the parts cannot be some separate mechanism inside the embryo that is not one of its parts. And so what makes the parts must be a part of the developing thing itself that is present in the seed (*sperma*) from the start (734a13–16, answering 733b32–734a1).[8]

742a16–b16 refers to some parts as *to gennêtikon* to indicate the role they play in generating other parts during morphogenesis. He does not mean those sexual parts that are involved in reproduction (e.g. testicles and penis).

8 Part of what makes the whole discussion difficult is that throughout *sperma* appears to be left ambiguous between the father's semen and the embryo (what he calls the *kuêma* at 734a5). On Aristotle's preferred theory, these are two separate entities. Indeed, it is crucial to his hylomorphic model of generation that no part of the male semen (*gonê*) becomes the matter for the thing it generates (e.g. *GA* I 21). However, at least some of Aristotle's rivals held the view that the male implants his seed into the womb which then grows into a new animal. On their theory there is no difference between the semen and the embryo; the semen becomes the embryo. I am inclined to

At this point the thesis on the table is that the first part of the offspring (e.g. its heart) is preformed inside the seed after having been fashioned inside the father. This part then becomes the agent that makes the rest of the parts. This is clearly only a working hypothesis. As we shall see, Aristotle eventually rejects this thesis at 734b17–19, where he concludes that there is something that makes the parts but it is not some determinate thing (*tode ti*) that is present in the embryo 'as something already complete from the start'. In other words, what makes the parts is not itself some other preformed part. However, at this stage in the argument Aristotle accepts that thesis as a working hypothesis and asks: 'Then how are *the other* parts formed?' (734a16). We are again given two options: either everything is formed straightaway at the beginning (full-blown preformationism); or the rest of the parts come to be in a sequence 'like the plaiting of a net' (partial epigenesis). He rejects the first option on empirical grounds. It is obvious to perception that the parts are not all there at the beginning; rather, some appear earlier and some later (734a20–5). This apparently shows that the parts come into being in a connected sequence (*ephexês*). But what kind of sequence is the process? Again, Aristotle canvasses two options: either it is a causal sequence where each part becomes a generator for the one after it (e.g. the heart generates the liver, which in turn generates some other part, and so forth in a series of discrete causal events); or it is merely a temporal sequence where each part comes into being *after* the other but not *by the agency of* the other. Aristotle rejects the first model on the grounds that it violates the causal principle that in all cases of coming-to-be the product *F* is generated out of something that is potentially *F* by something that is already *F* in a state of completion. If each part were to become the generator for the one after it, then the shape and form of the posterior part must already pre-exist in the prior part (e.g. the shape and form of the liver would already have to be in the heart), which Aristotle says is 'absurd and fantastic' (*atopos kai plasmatias*) (734a29–33). The implicit conclusion is that development forms a temporal sequence where the parts come into being after one another but not by the agency of

think that the current argument allows for this and does not depend on Aristotle's own technical theory of *sperma*. That is why he is not careful to make a sharp distinction between semen and embryo (see 734a35 where he says 'seed [*sperma*] or semen [*gonê*]' without any apparent distinction). This would also explain why, at this point in the argument, Aristotle proceeds on the assumption that the first part of the offspring is carried inside the seed after having been fashioned directly by the father. This is clearly only meant to be an interim conclusion that is accepted for the sake of argument, since it is eventually rejected.

one another. I shall return to the significance of this conclusion in a moment.

Finally, Aristotle rejects the working hypothesis (established at 734a13–16) that what makes the parts is itself a part of the developing thing that is present in the seed from the start: 'If everything comes to be from seed or semen [sc. 724a17–20], then it is impossible for some part of the animal or plant that comes into being to be present in the seed already from the outset'. For example, if the heart were fashioned directly by the father and then carried into the mother inside his seed, that part would not come to be *from* seed. If so, 'then it is impossible for some part of the animal or plant that comes into being to be present in the seed already from the outset'. For *ex hypothesi* all the parts come to be *from* seed (734a34–b4).

Herein lies the *aporia*. Aristotle starts from the premise that generation either has an external efficient cause (the father) or the efficient cause is something present in the embryo itself. But the argument shows that it cannot be either: the efficient cause of the animal and its parts cannot be anything present in the embryo; but it cannot be anything external to it either; and yet, it has to be one or the other.

Aristotle proposes to 'loosen' the puzzle by targeting the contact proviso underlying the argument against taking the efficient cause to be something external to the embryo. Perhaps there is a sense in which it is possible for the parts to be formed by an external agent. To demonstrate this, Aristotle appeals to the model of a self-moving automaton (734b4–17). In one sense, the parts of the automaton are moved by the external agent who triggers the mechanism (the primary mover). Yet, in another sense the parts are moved by the motion of the gears inside the machine itself (the proximate mover). Likewise, Aristotle says, in one sense the father makes the parts of the developing offspring, not by being in contact with them as they develop, but by having previously been in contact with the seed initially. Yet, in another sense, what makes the parts is 'the internal motion' (*hê enousa kinesis*) that is set going by the father just as the process of house-building makes the house (734b13–17). As I read this passage, 'the internal motion' refers to the process of development itself. This, I take it, is the force of the comparison with the process of house-building (the building motion is internal to the house being built just as the motions and changes that make up development take place in the animal that is coming to be).[9]

[9] cf. *GA* 730b7–8: *he oikodomêsis en tois oikodomoumenois*. The reference to an internal motion is meant to pick up the point from 734b7–8: 'It makes no difference whether we say the seed (*sperma*) or that from which the seed comes, in so far as the seed has within itself the movement initiated by the other.'

If this is right, then what Aristotle is saying here is that the developmental process constitutes the *proximate* efficient cause of the animal and its parts while the *primary* efficient cause is the father who first sets that cascading sequence of changes in motion. I suspect that this is why Aristotle says in the opening passage of the *GA* that 'to inquire into this [sc. the moving cause] and to inquire into the manner of generation for each thing is, in a way, the same thing' (715a13–15). For what we are doing when we are inquiring into the manner of a thing's generation is effectively inquiring into its proximate efficient cause.

This way of conceptualising the efficient cause of animal generation satisfies the causal principle that in all cases of coming-to-be the product *F* is generated *out of* something that is potentially *F by* something that is already *F* in a state of completion. (Aristotle takes this principle as his 'fresh starting point' at *GA* 734b20–2.) As we have seen, Aristotle rejects the version of epigenesis where each part of the animal is generated by the one that comes before it on the grounds that it violates this principle. On Aristotle's preferred model, what makes the parts is a single process that originates in the generating parent who is already in a state of completion what the matter from the female is in potentiality. This avoids the absurd consequences of that other model, because it treats the development of the parts not as a series of discrete generations (e.g. the heart generates the liver, the liver generates something else, and so forth) but as continuous stages in the generation of a whole animal (understood as a single causal event) whose generator is another substance that already bears the form of the product in a state of completion (*GA* 734b34–6, 735a3–4).

So far, the *GA* appears to support the Kelsey–Stavrianeas position that when Aristotle defines nature as an inner principle of change he does not mean an internal efficient cause. Even in the case of animal generation, while the *proximate* cause is in the developing organism, the *primary* efficient cause is the nature in the generating parent which is external to it. But then it is hard to see how animal generation is supposed to differ from craftsmanship. A house is also proximately caused by an internal process (734b16–17) whose origin is a principle of change in something other than the materials that undergo that change (the craft in the builder) (*GA* 734b33–735a4). But this is not Aristotle's final answer to the question

Peck translates *sperma* as 'semen' here. I prefer the more ambiguous 'seed' since I have suggested that Aristotle's use of *sperma* in this argument does not rely on his own technical distinction between the father's semen and the embryo whose development it sets going (see note 8). Nor is that distinction relevant here. The point Aristotle is trying to make is simply that *the motion itself*, which is set going by the father, constitutes the proximate efficient cause of the animal and its parts.

about the efficient cause of animals and their parts.[10] At the end of *GA* I Aristotle proposes the following solution to this 'very great puzzle':

> What is responsible for generation (*aition tês geneseôs*) is not a part of this but the primary external mover (*to prôton kinêsan exôthen*). For while nothing generates itself (*outhen gar auto heauto genna*), once it has been generated it makes itself grow (*auxei*). This is why a certain part comes to be first and not all of them together. And the part that comes to be first must necessarily contain the source of growth. Whether it is a plant or an animal, this is the nutritive principle that belongs to them all alike (which is also the capacity to reproduce another like oneself; for this is a function that belongs to all animals and plants that are naturally complete). The reason this is necessary is that once it has come to be, it must necessarily grow. And while it is begotten by something of the same name, e.g. a human by a human, it grows by means of itself. And since it makes itself grow, it must be an existing thing. And if, indeed, it is some one thing, and if this is first, then it must be the first part to come to be. Therefore, if the heart is the first to be formed in certain animals (and the part analogous to this in those that do not have a heart), the principle must come from this in those that have one (or from the analogous part in others). (*GA* II 1, 735a12–26)

In the context of living things *auxêsis* ('growth') typically refers to the process of structured growth that occurs after all of the organism's parts have been fashioned (see *GC* I 5). As the thing grows, its parts maintain their shape and form and simply increase in size (a change in the category of quantity). Understood in this sense, the above passage would seem to echo the claim Aristotle makes in *DA* II 4 when he says that the nutritive soul is 'the agent that effects generation (*geneseôs poiêtikon*), not of the very thing that is nourished, but something else like it; for its substance already exists, and nothing generates itself but only preserves itself'. I mentioned earlier that there is a tendency to read this doctrine as saying that the nutritive soul is a unified power to produce a certain substantial form this power *maintains* the organism's own form when directed on his own blood (nutrition) while it *reproduces* that same form in another individual when directed onto the menstrual blood supplied by the female (generation). When read in the context of that doctrine, it is very tempting to read the *GA* II 1 claim that 'nothing generates itself' (*outhen gar auto heauto genna*) but that 'once it has been generated it makes itself grow' (*auxei*) as claiming that the power of the offspring's own nutritive soul is only responsible for adding bulk to its already constituted parts.

[10] The following interpretation is also articulated by Johansen (2012, chapters 5 to 7) and carefully defended by Connell (2016, chapters 4 and 5).

But the argument itself suggests that *genesis* should be taken in a more restrictive sense to refer not to the whole generative process but to the conception of the embryo – or, more specifically, to the moment when the embryonic heart has appeared and begins to process nutriment. Aristotle draws special attention to this moment in *HA* IV 3 in his empirical description of bird development:

> With the common hen after three days and three nights there is the first sign of the embryo; with larger birds the interval being longer, with smaller birds shorter. Meanwhile the yolk comes into being, rising towards the sharp end where the origin (*archê*) of the egg is situated, and where the egg gets hatched. And the heart appears like a speck of blood in the white of the egg. This point beats and moves as though alive (*empsuchon*), and from it as it grows two vein-ducts with blood in them advance in a convoluted course towards each end of the two circumjacent integuments. And a membrane carrying bloody fibres now envelops the white, leading off from the vein-ducts. A little afterwards the body is differentiated, at first very small and white. (*HA* VI 3, 561a6–18, Thompson translation with slight modifications)

The suggestion I am making is that when Aristotle says that 'nothing generates itself', *genesis* is restricted to embryogenesis (up to that stage in the process where the embryo first appears as 'a speck of blood' that 'beats and moves as though alive'). Aristotle no doubt interprets the motion of that tiny speck as the embryonic heart beginning to process nutriment on its own. At that moment, the new individual can be said to *have come to be* ('since it makes itself grow, it must be an existing thing').[11] When Aristotle then says that 'once it has been generated it makes itself grow', *auxêsis* can be taken to refer to the stage where the rest of the body is first differentiated (morphogenesis).[12] That is why Aristotle says that the part that comes to be first must contain the source of this change. If Aristotle meant that the offspring makes itself grow only in the sense of adding bulk to its parts after they have been formed, then it wouldn't be necessary for any particular part to contain the source of that change. The reason Aristotle says the first part to emerge must contain that principle is that at that point the embryo must be ready to take over the construction of the rest of its body. If this

[11] As we shall see just below, Aristotle makes this same point at *GA* 740a18–20: 'For this reason, it is the heart that appears marked out first in all blooded animals; for this is the source (*archê*) of both the uniform and non-uniform parts. And this deserves to be called "the source" of the animal and the whole complex of parts (*tou sustêmatos*) from the moment it requires nutriment; for what exists grows.'

[12] There is precedent for this use of *auxêsis* at *HA* 561a14, where the term is used in connection with development ('as it [sc. the embryo] grows (*auxanomenou*), two vein-ducts with blood in them trend in a convoluted course towards each of the two circumjacent integuments').

interpretation is right, then *GA* 735a12–26 actually shows that the lion's share of generation is efficiently caused by an active power *in* the embryo itself (its nature or soul).[13]

This interpretation gains support from another argument in *GA* II 4 where Aristotle again tries to establish his thesis that the heart is the first part of the animal to come into being:

> This is clear not only to the senses (for it occurs in this way) but also according to reason. For when the thing generated has been separated (*apokrithê*) from both its parents, it must be able to manage itself (*auto hauto dioikein*) just as a son who has been sent away from his father. Therefore, it must contain a principle from which comes the subsequent ordering of its body (*husteron hê diakosmêsis tou sômatos*) in the case of animals . . . For this reason, it is the heart that appears marked out first in all blooded animals; for this is the source (*archê*) of both the uniform and non-uniform parts. And this deserves to be called "the source" of the animal and the whole complex of parts (*tou sustêmatos*) from the moment it requires nutriment; for what exists grows. (*GA* II 4, 740a4–20)

The key premise in this argument is the claim that as soon as the offspring has been 'separated' from both of its parents 'it must be able to manage itself just like a son who has been sent away from his father' (740a5–6). It is possible that Aristotle is thinking here of the fully formed baby that has become physically separated at birth. In that case Aristotle is making the obvious point that the newborn must in some sense be capable of feeding itself (to nourish or feed is one meaning of the Greek *dioikein*) since it is no longer relying on food processed by the mother or on the supply of yolk in the egg. But there are three reasons why the argument should not be read in this way.

First, Aristotle will go on to use this as a reason for claiming that the first part to come into being must contain a principle from which the subsequent *ordering* of its parts is derived (*hê diakosmêsis*) (740a8–9). This suggests that *dioiken* (a6) is understood as being able to manage or control morphogenesis. Second, at this stage in the *GA* Aristotle is still focused on what happens to the embryo at the very beginning of its development before any of its major structures have been differentiated out of it (see *HA* 561a6–18 above). This is clear from the immediate context. Just before this passage Aristotle characterises the embryo as a kind of plant seed that sends out roots from its primitive heart (cf. *HA* 561a14) in order to secure from the mother the matter necessary for building the rest of its body

[13] Connell (2016, e.g. 208).

(739b33–7). So, it is unlikely that the ensuing argument would have anything to do with what happens after parturition.[14]

Finally, if Aristotle's point was simply that babies need to be able to feed themselves in some sense once they are outside the womb or egg, then the argument would not support the conclusion that the heart is *the first part* to come into being in actuality, which is precisely what it is meant to establish (740a3–5). That inference clearly depends on the assumption that the development of the remaining parts is derived from a principle located in the embryonic heart. This is the force of the political analogy. As a young boy a son lives under the care of his father, who manages the family in virtue of possessing the craft of household management (*oikonomia*). After the son leaves the care of his father, he must be able to set up and run his own family (and not just feed himself). Before he can do that, he must acquire the craft for himself. Likewise in the case of animals: while the formation of the embryo is managed by motions in the father's semen which ultimately derive from his own nature (cf. *GA* 738b11–18, 743a34), once the embryo has been constituted it must be able to set up order in its own house as it were. This is why Aristotle says that it 'must contain a principle from which comes the subsequent ordering of its body'. And since this principle is located in the heart, that part must come into being before any of the others.

All of this suggests that by being 'separate' from its parents Aristotle is talking about the moment when the embryo becomes a *separate entity* with an independent ontological existence of its own (a *tode ti*).[15] This is the moment when the embryo is first constituted and begins to process nutriment through its newly minted heart. Before the embryo has been set (739b33), there is no identifiable individual separate from the parents' two reproductive fluids that are mixing together. Once it has been constituted and begins to process nutriment, it becomes a quasi-autonomous agent that is ontologically distinct from its parents. At that point the offspring can be said to exist as a separate individual in its own right ('for what exists grows').[16]

[14] See *GA* 739b33, which begins: 'Once the *kuêma* has been set (*sustê*)'. Aristotle consistently uses *kuêma* for the incomplete embryo. I am not aware of any place in the *GA* where *kuêma* is used to refer to the growing fetus that already has its parts fully differentiated, let alone for the newborn baby.

[15] Stavrianeas (2015, 59), Connell (2016, 146).

[16] The embryo obviously does not constitute a substance in a state of completion. In *PA* I 1 Aristotle suggests that it is ontologically more akin to a process than a substance: 'while the embryo (*sperma*) is a process of coming-to-be (*genesis*), the end [sc. the mature organism] is a substantial being (*ousia*)' (641b31–2). However, for the purpose of generation Aristotle considers the embryo to be a separate

We are now in a position to answer the opening question in *GA* I 1 about the efficient cause of the animal and its parts (715a13–15). The proximate efficient cause is the process of development itself (the 'internal motion'), which is initiated and controlled by a principle in the embryo itself (its own nature). This is the primary *internal* efficient cause. The father is the primary *external* efficient cause (*to prôton kinêsan exôthen*) that is responsible for forming the embryo and fashioning its primitive heart (embryogenesis). Once the embryo is constituted and begins to process nutriment, the new substance can be said to exist as a separate individual with an independent ontological existence of its own. For this reason the father can also be considered the primary efficient cause of substantial generation itself, which satisfies the causal principle at *GA* 734a29–31 (cf. 734b20–4, b34–6, 735a3–4). It should be clear by now how animal generation differs from craftsmanship. In both cases the proximate efficient cause of the product is a formative process that is ultimately derived from a potential for form located in the external agent (the nature in the father and the art in the craftsman, respectively). But in the case of biological generation, the potential for form is transmitted to the subject that is coming to be which allows it to take control of its own production. Thus, unlike artefacts, biological substances are self-generating systems (at least in the Aristotelian sense).

If this is right, then the explanatory resources of *DA* II 4 are too blunt to capture the nuances of Aristotle's mature theory of animal generation. On the standard picture of *DA* II 4, the nutritive soul is treated as a single capacity to (re)produce a particular substantial form. When operating on the father's own blood it maintains that form in his body (nutrition). When redirected onto menstrual blood it reconstitutes that same form in a new body (generation). But the *GA* picture is not as tidy as that.[17] According to the *GA*, the father's direct contribution to generation is limited to 'setting' the embryo and forming its primitive heart (embryogenesis). The subsequent development and organisation of the offspring's new body (morphogenesis) is controlled by the formal nature in the embryo itself, which serves as an *internal* efficient cause of that change just as the traditional reading of *Physics* II 1 maintains.[18]

quasi-autonomous individual from the moment it has been constituted and begins to process nourishment.

[17] This is the sense of the *GA* one gets from Menn (2002). Admittedly, there are some passages in the *GA* that do suggest this picture (e.g. *GA* 730b8–32, 737a18–22). But these texts are rare and obscure the more complex account of generation developed in that treatise (see below).

[18] This does not mean the *GA* is incompatible with the *DA* II 4 doctrine. Aristotle warns us at the end of *DA* II 4 that what has been said there is only an 'outline' (*tupos*) to be worked out later in the

7.4 The Mechanisms of Heredity

I now want to take a closer look at how the organism's potential for form (its nature or soul) is transmitted from parent to child in the act of reproduction. To do this we need to turn to Aristotle's theory of heredity in *GA* IV 3. This will also help fill in the details of the view in Chapter 6 that what the generator imparts to the matter supplied by the female is not a latent soul but certain active powers (*dunameis*) which, in conjunction with the matter's own passive powers, cause it to develop the sorts of organs whose own developed capacities ('first-actuality') constitute its soul.[19]

Aristotle sets out the basic mechanisms of heredity in a key passage towards the beginning of the chapter:

> I speak of each *dunamis* in the following sense. The generator is not only a male but also a particular sort of male, e.g. a Coriscus or a Socrates, and it is not only a Coriscus but also a human being. And it is in this sense that, of the things that belong to the generator insofar as it is capable of generating (*huparchei tôi gennôti katho gennêtikon*) and not incidentally (e.g. if he is a scholar or someone's neighbour), some belong to it more closely while others more remotely. That which is distinctive (*to idion*)[20] and particular (*to kath' hekaston*) always exerts a stronger influence relative to generation. For Coriscus is both a human being and an animal, but human is closer to what is distinctive of him than animal. And while both the particular and the kind generate, it is more so the particular; for that is the substance (*hê ousia*). For while the offspring that comes to be becomes a certain sort of thing (*poion ti*), at the same time it comes to be a determinate individual (*tode ti*), and this is the substance. So there are motions (*kinêseis*) present in the parent's seed derived from the potentials (*dunameis*) of all of these sorts of properties, and in potentiality even those of its ancestors, although those of the individual are always closer. (*GA* IV 3, 767b23–768a2)

Aristotle does two things in this passage. He first distinguishes between those features that belong to the generator 'insofar as it is capable of generation' (*katho gennêtikon*) and those that are incidental to it as a generator. He then goes on to introduce the causal mechanism that underwrites the transmission of those former attributes. The two central components of the mechanism are the "motions" that are said to be present

appropriate treatises (*oikeiois logois*). This is likely a reference to the *GA* (and possibly the lost work, *On Nutrition*).

[19] The following is based on Henry (2006a), which is followed closely by Gelber (2010) and Connell (2016, 172–7).

[20] See *GA* II 3, 736b4–5: 'For the end is formed later in the process, and what is distinctive (*to idion*) to each generated thing is an end.'

in the spermatic material of all sexually reproducing animals and the "powers" or "potentials" from which those motions are derived, each of which is said to correspond to one of those reproducible features mentioned in the first part. I begin with the first distinction.

The heritable–incidental distinction set out in the first part of this text is clearly meant to bring to bear Aristotle's more general doctrine of intrinsic (*kath' hauto*) causation on the theory of reproduction.[21] According to that doctrine, we say that *A* produces *B* in virtue of itself and non-incidentally if *B* results from *A* exercising a causal power for *B*. If, in exercising that power, *A* also happens to produce something else *C*, then we say that *A* was an intrinsic cause of *B* but only an incidental cause of *C* because *A* was not exercising a power for producing *C*. Aristotle offers the following example in *Metaphysics* VI 2:

> A confectioner, while targeting (*stochazomenos*) pleasure, may make something wholesome, but not in virtue of the confectioner's craft. For this reason, we say it was an accident (*sunebê*) and that in a way he makes something wholesome, but he does not make it in the unqualified sense [sc. in virtue of his craft]. For the producers of some incidental results are other powers (*allôn dunameis*), while in other cases there is no definite craft or power. (*Metaph.* 1027a3–7)

The confectioner's craft is a power for producing sweet things that are pleasant to eat. If, in exercising that power, she also happens to produce something wholesome, then we say that *in a way* she made something wholesome but only accidentally since she did not produce it in virtue of her craft. The fact that the outcome was something wholesome was merely an incidental byproduct of the activation of her power to produce something pleasant.

GA 767b23–768a2 applies this doctrine of intrinsic causation to the theory of reproduction. In the closing lines of the passage Aristotle says there are motions present in the parent's seed 'derived from the *dunameis* of all of these sorts of things', where 'all these sorts of things' picks up the earlier reference to those properties that belong to the generator in virtue of itself (*katho gennêtikon*) and not accidentally (*kata sumbebêkos*), viz. those that make it male, Socrates, and human. Notice the distinction Aristotle draws here between a property of the generator (e.g. Socrates' actual snub-nose) and the *dunamis for* that property. Let us assume that the concept of *dunamis* is being used here in the familiar Aristotelian sense of a causal power

[21] Meyer (1992) gives a good overview of that doctrine.

(potential, capacity). What exactly are these causal powers? Recall that back in *GA* II 4 Aristotle had said that the productive power (*hê poiousa dunamis*) that constructs the organism at the beginning belongs to the same part of the soul that is responsible for its growth and nutrition later on: 'If, therefore, this is the nutritive soul, then it is also the power that generates. And this is the nature of each thing, which is present in all plants and animals alike.'[22] What I want to suggest is that the *dunameis* identified in our *GA* IV 3 passage are meant to pick out different aspects of this generative power – what Allan Gotthelf calls the organism's 'single potential for form'.[23] If this inference is correct, then Aristotle's hereditary concept of *dunamis* can be seen as providing the ontological basis for the heritable–incidental distinction set out in the first part of this text.

What makes it true to say that Socrates is an *intrinsic* cause of the generation of a snub-nosed human male but only an *incidental* cause of a good flute player or someone's next door neighbour is that the former outcome can be traced to a set of active powers in Socrates' formal nature that reliably produce that effect when operating on a suitable material with the right passive powers (viz. human blood). If, in exercising those same powers, Socrates also happens to generate a good flute player or someone's next door neighbour, then we can say that *in a way* he is a cause of those other things but only accidentally since they did not result from of any causal powers for those effects (*Metaph.* 1027a6–7: 'for the producers of some incidental results are other powers').

While *GA* IV 3 is aimed primarily at explaining the capacity to generate certain family resemblances (767a36–b5), Aristotle extends the mechanism to the inheritance of those universal properties that belong in common to all members of the species as well as those that are common to all animals (e.g. 767b23–768a2, 781a8–12). In this way *GA* IV 3 can be seen as deploying the general doctrine of intrinsic efficient causation to explain how all heritable features of an organism's body

[22] See *GA* 740b25–741a3, translated above.

[23] By a 'potential for form' Gotthelf means an active power (*dunamis*) for generating the form of the mature organism. Both Gotthelf (2012, e.g. 100) and Leunissen (2010, e.g. 208–9) associate this potential for form with the organism's formal nature in its role as a goal-directed efficient cause. In *DA* II 4 Aristotle also defends the thesis that the power to generate the parts at the beginning (morphogenesis) and the power to grow and nourish those parts later on (nutrition) belong to the same part of the soul. When Aristotle tells us in *GA* II 5 that the nutritive powers of the offspring's soul come from the mother, I take it he has in mind the latter function. By contrast, my suggestion is that the *dunameis* in our passage from *GA* IV 3 are active powers associated with the generative function of the nutritive part.

are transmitted from one generation to the next in the act of reproduction.[24]

The other part of the mechanism concerns the motions that are carried inside the parent's seed which are said to be 'derived from' the *dunameis* for its heritable features (767b36, 768a11–14, 768b4–8). Above I argued that these *dunameis* are meant to pick out different aspects of the generator's potential for form (its formal nature). I now want to suggest that the job of the spermatic motions is to transmit this potential for form and that the way they do this is by passing on to the embryo a set of active powers like those from which they were originally drawn.[25] Later on in *GA* IV 3 Aristotle makes two important modifications to this theory. First, he posits a set of spermatic motions coming from the female herself (768a14–21). The most natural reading of this passage is to take those motions as functionally equivalent to the ones coming from the male.[26] This means that the female also contributes a set of 'formative motions' (768a15–16: *hai kinêseis hai dêmiourgousai*) which – like those in the semen – ground the menstrual blood's ability to actively transmit resemblances from her and her ancestors. Second, although Aristotle first elaborates the theory by referring to motions and potentials for whole organisms (e.g. 'the motion coming from Socrates'), he eventually extends the mechanism to include motions for each part of the organism (768b1–5). This allows him to explain how the offspring can resemble different individuals with respect to different parts (cf. 767b1).

The *GA* IV 3 account of heredity also provides Aristotle with a causal mechanism for explaining how the parents can each make separate contributions to the offspring's soul without committing himself to the view that the parts of the soul exist in separation from one another inside the parents' seeds (see Chapter 6, Section 6.3). On the interpretation developed here, what the father transmits in his semen are a series of motions whose function is to endow the embryo with a set of active powers (*dunameis*) for forming sensory organs, while those from the mother

[24] Witt (1985, 52–4) argues that *GA* IV 3 is limited to the transmission of family resemblances. But Aristotle explicitly extends the causal mechanism to universal features that belong to the offspring as a human and an animal at 767b23–768a2 and 781a8–12.

[25] For a similar distinction between the *kinêseis* and the *dunameis* in *GA* IV 3 see Connell (2016, section 9.5). Aristotle's picture is more complex than this. Each parent transmits its own set of *kinêseis* drawn from a corresponding set of *dunameis* from its own nutritive soul. The complicated 'laws' developed in *GA* IV are supposed to govern the interactions between these various motions in a way that determines the particular pattern of heredity for the offspring. See Henry (2006a).

[26] For an extended defence of this point see Henry (2006b). A similar conclusion is reached by Connell (2016, chapter 9), albeit through a different line of reasoning.

include active powers for developing parts of a nutritive system. The developed capacities (or first-actualities) of those different physiological systems constitute the sensory and nutritive parts of the offspring's soul, respectively.

7.5 Two Senses of Formal Nature

In this chapter I have drawn attention to Aristotle's definition of nature in *Physics* II 1 as 'a principle or cause of being moved and of being at rest in that to which it belongs primarily, in virtue of itself, and non-accidentally' (192b20–3). I have tried to show how the evidence from the biological works might support the traditional view that Aristotle is defining nature here as an internal *efficient* cause.[27] But there is a sense in which the Kelsey–Stavrianeas interpretation is correct insofar as nature also serves as an internal *final* cause. In *Physics* II 2 Aristotle goes on to say that natural substances have two natures (*hê phusis dichôs*, 194a12; *duo hai phusei*, 194a16), a material nature and a formal nature, and that it is up to the natural scientist to know both of these up to a point (194a12–27). Briefly, the material nature of a thing includes all of the powers and capacities rooted in its basic physical make-up, while the formal nature (at least in the case of living things) is constituted by the set of capacities it has for executing various life-functions (its soul). Both of these count as natures because both the capacities of a thing's physical make-up and its capacities of soul are the intrinsic sources of the characteristic changes undergone by that substance.[28] Finally, in *PA* I 1 Aristotle makes a further division between two senses of formal nature based on how each serves as a principle of change:

> The natural scientist will state both what the soul (or that very part of the soul) is and speak about the attributes it has in virtue of the sort of being it is, especially since the nature of something is spoken of and is in two ways, namely, as matter (*hôs hulês*) and as substantial being (*hôs ousias*). And nature understood as substantial being encompasses both nature as mover (*hôs hê kinousa*) and nature as end (*hôs to telos*). And it is the soul – either the whole soul or some part of it – that is of this sort in the case of animals.[29] So in this

[27] There is some debate over how this applies to the elements in light of Aristotle's remarks in *Ph.* VIII 4. My own sympathies lie with Shields (2007).

[28] For a good discussion of this point see Cooper (1982). Even if it turns out that matter qua matter is entirely passive and inert (see Chapter 4, Section 4.4), matter still counts as a nature insofar as it is a source of being *passively* changed in that to which it belongs primarily and in virtue of itself.

[29] Aristotle seems to have in mind here the fact that the rational part of the human soul is not a 'nature' insofar as it is not a source of change (641a5–9). The reason is that change requires a material substratum and thinking is not an enmattered capacity (cf. *DA* II 1, 413a3–10; *DA* II 5; *GA* II 3, 736b23–737a16).

way it will be requisite for the student of nature to speak about the soul more than the matter, inasmuch as it is more that the matter is a nature because of the soul than the reverse; for indeed the wood is a bed or a stool because it is potentially these things. (*PA* I 1, 641a22–33, Lennox translation with modifications)

It is not clear from this passage how Aristotle thinks these two senses of formal nature ('nature understood as substantial being') are related to one another in a living substance. At the very least we can draw a conceptual distinction between them owing to their different causal roles in the generation of a new substantial individual.[30] For example, at the end of *Metaphysics* 1032a12–25 (translated above) Aristotle tells us that that *by which* a natural substance comes into being is what we call the formal nature (*hê kata to eidos phusis*). This is what our *PA* passage calls formal nature 'as mover'. But the concept of nature can also be applied to the mature form that defines the substance and specifies the goal of its development (*Metaph.* V 4, 1015a10–11; *Ph.* II 2, 194a28–9). This corresponds to nature 'as end'.[31] Since what it is to be an efficient cause of change differs from what it is to be the final cause of that change, it follows that nature 'as mover' and nature 'as end' are at least conceptually distinct.[32] In the next chapter I examine how the three senses of nature (nature as matter, nature as mover, and nature as end) causally interact in the dynamic teleological context of morphogenesis.

[30] Contrast Broadie [Waterlow] (1982, 64–5), who claims that there is a 'conceptual coincidence' that obtains between these two senses of formal nature.

[31] Compare *GC* II 9, 335b4–6 (Chapter 4) where Aristotle says that the form of a thing is both 'a cause in the sense of that for the sake of which generation takes place' and 'the account of the substance of each thing'. As I read this, Aristotle means that this one ontological item – the form – plays two different causal roles depending on the context. It is both the formal cause of being an *F* (the form is the feature that defines the substance and thus causes it to be an *F*) and the final cause of becoming an *F* (it is also the goal towards which its generation is directed and explains why it undergoes the transformations it does). Compare *DA* II 4, 415b8–27, where the soul is also said to be a formal, final, and efficient cause of the living body (cf. *Ph.* II 7, 198a25–7).

[32] Cf. *GC* I 7, 324b13–18 and Lennox (2001a, 189).

CHAPTER 8

The Architectonic Model

The seeds of an answer to the question raised at the end of Chapter 7 – how do formal and material natures interact in the creation of a new substantial individual? – have already been sown in Chapter 4 in connection with *GC* II 9. On the reading I defended there, while Aristotle agrees with those more sophisticated materialists who regard natural generation as the effect of the simple bodies and their inherent powers, he ultimately rejects their account on the grounds that the powers they give them are too instrumental (*lian organikas*) to serve as primary causes of the change (336a1–12). But Aristotle's positive account in *GC* II 9 remains undeveloped in two important respects. First, he says very little about the nature of the efficient cause that was 'vaguely dreamed of' by all his predecessors but 'definitely stated' by none of them (335b7–8). While I argued that what Aristotle is after in *GC* II 9 is the sort of efficient cause that can explain the order, beauty, and goodness of natural generation, Aristotle doesn't make that point explicit in *GC* II 9. Instead, I inferred it from the parallel discussion in *Metaphysics* I 3. There Aristotle argues that even if we accept that some of the elements are endowed with motive natures of their own, they cannot explain how natural things come to be 'in a good and beautiful condition' (984b8–22).[1] Second, *GC* II 9 says very little about how this sort of efficient cause interacts with the simple bodies and their inherent powers, which the biological works agree are at work in the

[1] I have argued that in the case of living things beauty and goodness are associated with their teleological structure (*PA* 645a24–6). In this context to say things come to be 'in a good and beautiful condition' is to draw attention to the fact that living things and their parts exhibit the sort of adaptive complexity that promote their survival and well-being. What is responsible for this is the formal nature of the organism operating in its role as a goal-directed efficient cause. In the case of the elements, their 'beauty' resides in the fact that their mutual transformations and the changes that follow upon them (e.g. annual rainfall) follow regular patterns as opposed to being random and haphazard. What is responsible for this order is the annual rotation of the sun in the ecliptic.

generation of a living thing. The aim of this chapter is to fill in that picture.

A place to begin is with a passage from *GA* IV 3 where Aristotle is discussing birth defects. While Aristotle often uses the model of desire to express the relation between matter and form (e.g. *Ph.* 192a20–5), in this passage he depicts their interaction using the more antagonistic model of political rule: birth defects are said to result whenever the formal nature 'fails to rule over' (*mê kratêsê*) its material nature (*GA* IV 3, 770b15–17).[2] What does it mean for one nature to 'rule over' or 'master' the other nature in the generation of a new substance? What is the essence of the rule that the formal nature fails to exercise over the material nature in this context? I shall argue that Aristotle was influenced here by the model of generation developed in Plato's *Timaeus* and that recognising this tells us something important about how he envisions the teleological control that is operative in the formation of a new substance. My argument runs as follows. In Section 8.1 I show how in the *Timaeus* Plato also describes generation in terms of Reason 'ruling over' Necessity by 'persuading' it to direct its activity towards a good end. I call this the architectonic model of generation because of the regulatory role it assigns to the primary efficient cause in the process. I then go on to argue in Sections 8.2 and 8.3 that this is the same role that Plato assigns to the *architectôn* (or master builder) at the beginning of the *Politicus*, which serves as a model for the kind of rule exhibited by those who practise the kingly art. This political conception of rule (I argue) serves as the basis for the teleological accounts of generation that we find in the *Timaeus*. Finally, in Section 8.4 I argue that Aristotle adapts this same architectonic model from the *Timaeus* as the basis for his own teleological account of morphogenesis in the *Generation of Animals*, while attempting to naturalise it by replacing talk of Divine Reason 'persuading' Necessity with talk of the formal nature of the animal 'using' the necessary motions of the simple bodies (e.g. heatings and coolings) as 'tools' for realising its good ends.[3]

[2] This use of *kratein* in embryology has its roots in both the Hippocratics and the Presocratics (see Connell 2016, s.v. 'mastery'). Although Aristotle uses this language most explicitly in connection with sex determination and inheritance, *GA* 770b15–17 seems to be expressing a more general idea about the relationship between the formal and material nature in the context of morphogenesis (the formation of the parts). The argument that follows does not turn on this claim.

[3] For an alternative (and rather uncharitable) account of the teleology of the *Timaeus* and Aristotle's attitude towards it see Bolton (2015, 126–8). For a discussion of the ways in which Aristotle naturalises Plato's teleology see Lennox (1999).

8.1 The *Timaeus* Model of Generation

In the *Timaeus* Plato represents the cosmos as an ordered whole that exhibits beauty and goodness and as such calls for an order-producing cause capable of generating those features (Johansen 2004, 72). Such a cause must be something that possesses intelligence and looks towards a form as its model, conditions that Plato thinks are satisfied by craftsmanship. The fact that the visible universe is the product of a Divine Craftsman who looked to eternal Forms as his models is thus supposed to explain why it exhibits beauty, order, and goodness:

> As I see it, then, we must begin by making the following distinction: What is that which always is and has no becoming, and what is that which is always coming to be and never is? The former is grasped by intelligence together with an account and has everlasting being, while the latter is grasped by belief together with non-rational sensation, and it comes to be and passes away but never really is. Now everything that comes into being necessarily comes to be by the agency of some cause (for it is impossible for anything to come to be apart from a cause). So, whenever the craftsman looks at what is always changeless and, using a thing of that kind as his model, reproduces its form and character, then, of necessity, all that he so completes is beautiful (*kalon*). But when he looks at a thing that has been begotten, his work will lack beauty. (27d5–28b4, translated after Zeyl)

> Now, surely it's clear to all that it was the eternal model he looked at, for, of all the things that have come to be, our universe is the most beautiful (*kallistos*), and of the causes the craftsman is the most excellent. This, then, is how it has come to be: it is a work of craft, modelled after that which is changeless and is grasped by a rational account, that is, by wisdom. (29a5–10, Zeyl translation)

In what follows I want to focus on two key methodological passages, *Timaeus* 46c7–e6 and 47e3–48a7. I begin with the former.

After giving an account of vision in terms of various elemental forces Plato has Timaeus make the following point:

> Now all of the above are among the co-causes (*tôn sunaitiôn*) employed in the service of the god as he attempts to bring to completion the character of what is best as far as it is possible. But because they make things cold or hot, compact or disperse them, and produce all sorts of similar effects, most people regard them not as co-causes, but as the [primary] causes of things. Things like these, however, are totally incapable of possessing any reason or intelligence about anything. We must pronounce the soul to be the only thing there is that properly possesses understanding. The soul is an invisible thing, whereas fire, water, earth, and air have all come to be as visible bodies.

> So anyone who is a lover of understanding and knowledge must necessarily pursue as primary causes those that belong to the intelligent nature, and as secondary all those belonging to things that are moved by others and that set still others in motion by necessity. We too, surely, must do likewise: we must describe both kinds of cause, distinguishing those which possess intelligence and thus fashion what is beautiful and good, from those which produce chance effects that lack order (*to tuchon atakton*) on each occasion when bereft of intelligence. (46c7–e6, Zeyl translation with modifications)

Plato makes two important distinctions in this passage. The first is between (1) those agents that possess reason and (2) those inanimate agents (the simple bodies and their inherent powers) that Plato had invoked previously in his account of the construction of the human body. (These latter are the sorts of agents that Plato associates with the Presocratic materialists at *Laws* 889b1–c6, each of which is said to be 'impelled by virtue of its own inherent properties' which in turn depend on 'various suitable amalgamations of hot and cold, dry and moist, soft and hard, and all other chance combinations that resulted from necessity when the opposites were mixed'.) Because these other agents lack soul, Plato says they must be 'totally incapable of possessing any reason or intelligence about anything' since reason can only come to be in the soul. At the end of the passage we are told that when these non-rational agents (e.g. fire, the hot and the cold, etc.) are operating on their own without being directed by some intelligent agent they produce 'chance effects that lack order'. So, we do not need to invoke reason to explain infrequent chance effects; blind mechanical forces are sufficient to explain those phenomena. Instead Plato thinks that intelligent causes are necessary to explain why the world is *beautiful and good*, by which he means the fact that things that come to be exhibit order, complexity, and the adjustment of means to ends (cf. *Gorgias* 503–506, *Laws* 889b1–e2, *Philebus* 30c3–6).

The other distinction Plato makes in this passage is a distinction between the different causal roles that these two types of agents play in coming-to-be. Reason is identified as the 'primary cause' (46d8: *aitias prôtas*) of the beauty and goodness of the visible world, while the elemental forces are downgraded to mere 'co-causes' (46d1). To call these other things 'co-causes', I take it, Plato means that, while they are in some sense responsible for generating the product, they ultimately owe their productive behaviour to some more fundamental cause (intelligence) that serves as the primary source of the change (see 68e1–69b2). Compare the role of the hammer in producing the sword. The pounding action of the hammer against the iron is directly responsible for shaping the blade in the sense that its sharp edge follows as a consequence of the hammer's blows; the hammer's blows are, as Aristotle would say, the proximate cause of the sharpness. But the motions of the hammer themselves are ultimately due to the blacksmith who uses the hammer as a 'co-cause' in bringing about that end.

At the beginning of our passage Plato says that because things like heating and cooling, compression and dilation, produce the sorts of effects we normally witness in coming-to-be the materialists had regarded them not as co-causes but as primary causes of things. I argued in Chapter 4 that this is essentially the same criticism Aristotle raises against the materialists at the end of *GC* II 9. By treating things like fire or the hot and the cold as primary causes of natural generation the materialists 'act in virtually the same way as one who treats the saw and the other tools of carpentry as the [primary] cause of the things that come to be' (336a7–11; cf. *GA* 789b3–15).

The other passage I want to pick up on comes at 47e3–48a7. In this passage Plato introduces three additional concepts that are central to the *Timaeus*' model of generation: necessity, persuasion, and the so-called straying cause.

> Now in all but a brief part of the discourse I have just completed I have presented what has been crafted by Reason. But I need to match this account by providing a comparable one concerning the things that have come about through Necessity (*ta di' anankês*). For this ordered world is of mixed birth: it is engendered from a union of Necessity and Reason. Reason ruled over (*archontos*) Necessity by persuading it (*peithein*) to lead most of the things that come to be towards what is best, and the result of Necessity's being overcome by wise persuasion was the initial formation of the universe. So if I'm to tell the story of how it really came to be in this way, I must also mix in the cause whose nature is to set things adrift (*to tês planômenês eidos aitias*). (47e3–48a7, translated after Zeyl)

The passage begins by saying that the universe is of mixed lineage; it is the offspring of a union of Necessity and Reason. As such, things that come to be require a twofold explanation: one given in terms of rational causes, the other in terms of necessity.[4] To explain something in terms of rational

[4] Compare 68e1–69b2, where Timaeus says of the simple bodies: 'And so all these things were taken in hand – their natures being determined then by necessity in the way we've described – by the craftsman of the most perfect and excellent among the things that come to be, at the time when he brought forth that self-sufficient, most perfect god. Although he made use of the relevant co-causes, it was he himself who gave their fair design to all that comes to be. That is why we must distinguish two forms of cause, the divine and the necessary.'

causes is to show how it was intelligently designed to serve a good purpose. Reference to that purpose ultimately explains why Reason designed the thing in the way that it did. For example, in the case of vision the god gave us eyes for the sake of observing the orderly revolutions of the heavens, which has the effect of stabilising the disorderly revolutions in the soul. Stabilising the motions in our soul is the reason for the sake of which we have eyes (47a1–c6).

There are good reasons to think that Plato means to associate the necessary causes with those elemental forces described as 'co-causes' in 46c7–e6 (the simple bodies and their inherent powers).[5] First, both the necessary causes and the co-causes are identified as the things that are directed by Reason towards what is best. Second, there are at least two places where the material agents that play the role of co-causes are associated with necessity in some way: at 46e1–2 they are said to be 'moved by others and set yet others in motion *of necessity*'; and at 68e1–69b2 Timaeus says that the simple bodies that serve as co-causes have their intrinsic properties and dispositions 'of necessity'. All of this indicates that when Plato refers to Necessity being 'persuaded' by Reason he is referring to those elemental forces operating in their role as co-causes (cf. 69a). The use of necessity as a co-cause of generation is a central feature of the architectonic model that I develop below.

It is important here that the type of necessity Plato has in mind is what Aristotle scholars call material or Democritean necessity and not what Aristotle calls conditional (or hypothetical) necessity.[6] According to this idea, 'material things have powers which determine, given the circumstances, how they will behave' (Manson 2006, 284–5). To explain some phenomenon in terms of material necessity involves showing how it follows from the physical natures of the materials and their causal interactions. For example, Timaeus says that fire is the most mobile of the elements because it is constituted from the tiniest solids with the fewest number of faces (pyramids) and that in every case 'the body that has the fewest faces is *of necessity* the most mobile, in that it, more than any other, has edges that are the sharpest and best fit for cutting in every direction' (56a3–b1). This illustrates the claim at 68e1–3 that the elements have their intrinsic properties and dispositions of necessity. In this case the direction

[5] Cf. Johansen (2004, 94–5). The pattern of explanation outlined in these two passages, where Necessity is used by Reason as a co-cause, is taken up primarily in the latter section of the *Timaeus* (69a–92b).

[6] Morrow (1950, 153), Manson (2006, 288–9), Johansen (2004, 94–9). For the concept of conditional necessity see *Ph.* II 9, *PA* I 1. See also (e.g.) Cooper (1987).

of necessity is bottom-up where properties existing at the lower level (having the fewest faces and sharpest edges) necessitate – in a non-teleological way – certain other properties and dispositions that emerge at the higher level (increased mobility). While it may be true that Reason intended to make fire the swiftest and most mobile of the elements and so selected solids with the fewest faces in order to realise that goal, given that fire has that material structure it is *necessary* that it be swift and mobile. Fire turns out to be the most mobile element simply in virtue of being composed of those kinds of solids regardless of whether or not it was designed with that purpose in mind. (It would still have that property even if it just happened to acquire that structure by chance.) It is the fact that the materials have an independent capacity to necessitate their effects in a reliable way that Plato thinks of them as being suitable to be used by Reason as 'co-causes'. This is the same as in craftsmanship. It is because of the necessary causal powers of steel that a hammer and anvil can be reliably used by the blacksmith as a co-cause in producing the sword.[7]

I will pass over the straying cause – 'the cause whose nature is to set things adrift' – since it is not directly relevant to the point I want to make in this chapter. I am sympathetic to Johansen's (2004, 94–5) view that this cause refers not to some independent principle over and above Reason and Necessity that is the source of indeterminacy in the world but to an additional way in which Necessity functions as a cause. On this reading co-cause and straying cause pick out the necessary causes operating in differ modes: "co-cause" refers to the role of those necessary physical forces when directed by intelligence towards an end; "straying cause" refers to those same necessary forces when operating on their own without intelligence, in which case Plato says they produce 'chance effects that lack order'.

Finally, Plato uses the idea of wise persuasion here as a metaphor for understanding the relation between Reason and Necessity and how the elemental forces at work in generation serve as co-causes. To explain what Plato has in mind I want to take somewhat of a detour into the *Politicus*, which I shall suggest is operating in the background of the *Timaeus*.

8.2 The Statesman's Rule in the *Politicus*

Commentators who seek to understand the use of the craft analogy in the *Timaeus* often turn their attention to the *Gorgias* where Plato develops the

[7] Johansen (2004, 98): 'the usefulness of the contributory causes as such requires that they necessitate their effects in a regular and predictable way.'

notion of a *technê* as an expertise that is oriented towards the good condition of its product, where the good of a thing is identified with some kind of order and arrangement among its parts (*Gorgias* 503–505). The connections between the *Timaeus* and Plato's model of craftsmanship are well-established, and I do not have anything to contribute to that debate. Without question the idea of a *technê* as an order-producing cause that operates by looking to a form as its model provides the general framework for the creation story developed in the *Timaeus*. For, as we said, the fact that the primary cause of the universe is a Divine Craftsman who arranged everything by looking to eternal Forms as his models is supposed to explain why the cosmos exhibits beauty, order, and goodness rather than their opposites. However, not much is made of the fact that in the *Timaeus* Plato uses the language of Reason 'ruling over' Necessity by means of 'persuasion', which suggests that the *Timaeus* is working with a distinctly political conception of generation. In light of this I want to suggest that when Plato speaks of the universe as the product of Reason what he has in mind is not so much ordinary craftsmanship as statecraft. I am thinking specifically about the conception of statesmanship introduced in the *Charmides* and *Euthydemus* and developed in the *Politicus*. Examining the account of the statesman's craft in the *Politicus*, I want to propose, can help shed light on how Plato understands the relation between Reason and Necessity in the creation of the universe and, indirectly, how Aristotle thinks about the relation between the formal nature and material nature in the development of a new organism.[8]

In the *Politicus* the Eleatic Stranger is engaged in a discussion with Socrates the Younger about the nature of statesmanship.[9] They begin by dividing knowledge into practical and theoretical knowledge and locate statesmanship on the theoretical side. They then divide theoretical knowledge into the sort that makes judgements and the kind that issues directives to others for the sake of coming-to-be (260a9–261b3). They follow the directive branch of theoretical knowledge until they end up with an account of the statesman as a kind of shepherd of the human herd (268c). This definition proves to be problematic because it fails to separate off statecraft from all those other crafts that claim to

[8] Morrow (1950, 162) also recognises the political tone of the *Timaeus* model but does not connect it to the *Politicus* in the way I do. Instead he looks forward to the *Critias*. One might object here that it is perfectly natural for Plato to describe ordinary craftsmanship as 'ruling over' his subject without invoking any political conception of rule, as he does at *Euthydemus* 291e4–292a2. But notice that in that passage talk of a craftsman 'ruling over' his subject is intended precisely to make a point about the kingly art (cf. *Republic* 342c8). So the context is overtly political.

[9] The following discussion has benefited greatly from the study of Lane (1998) and Gill (2012, chapter 6).

look after various aspects of human life. At this point the Stranger introduces a myth which is meant to expose the errors of the first division, most importantly the fact that they failed to specify the manner of the statesman's rule (274e6–275a10). Eventually he offers the example of weaving as a model for how to differentiate among a set of related crafts all of which deal with the same object. For like statesmanship there are countless other crafts that share with the weaver the task of looking after and producing clothes (281b8–11). The main contribution of the model is the introduction of the distinction between 'cause' and 'co-cause' (Lane 1998, 58). For example, weaving can be separated off from the expertise of the shuttle-maker insofar as the latter is only a co-cause in producing clothing insofar as it provides the weaver with the tools necessary for doing his job (281c2–e10). In a similar way, the other crafts that rival the statesman for the title of caretaker of the human herd are separated off as co-causes that provide the necessary tools and materials for accomplishing his ends (287b4–8). While this significantly advances the discussion, the Stranger notes that it fails to differentiate the kingly art from a final group of closely related crafts that cooperate in 'jointly governing the affairs of the city' (*sundiakuberna tas en tais polesi praxeis*, 304a2–3): generalship, rhetoric, and the art of the judge (303d).

At this point it becomes necessary to return to the first division and reconsider the claim that statecraft is a purely theoretical kind of knowledge. Back at 259c7 the Stranger had said: 'But this much is clear, that the power of any king to maintain his rule has little to do with the use of his hands or his body in general in comparison with the understanding and force of his mind.' This is not entirely true, as Plato reveals towards the end of the dialogue. While it is true that the statesman's expertise turns out to have little to do with the use of *his* hands or *his* body, he is not completely detached from the practical affairs of the city. The initial division failed to accommodate this practical dimension of the statesman's expertise. Plato had already foreshadowed the solution to differentiating statecraft from the other subordinate crafts at 259e11 where the Stranger offers Young Socrates the example of the *architektôn* (or master builder) who supervises the productive activities of the workers: 'And all master builders too, they don't act as workers themselves, but manage workers ... Insofar as what the master builder provides is understanding rather than manual labour.' Towards the end of the dialogue, at 305c–e, it is revealed that this architectonic feature is part of the essence of statecraft:

> If then one looks at all the sorts of expert knowledge that have been discussed, it must be observed that none of them has been declared to be statesmanship. For true kingship must not perform practical tasks (*prattein*) itself but control [or rule over, *archein*] those with the capacity to perform them because it knows when it is the right time to begin and set in motion the most important things in cities and when it is the wrong time. And the others must do what has been prescribed for them ... For this reason, then, the sorts of expertise we have just examined [sc. generalship, rhetoric, art of the judge] control neither each other nor themselves, but each is concerned with some individual practical activity of its own, and in accordance with the individual nature of the activities in question has appropriately acquired a name that is individual to it ... Whereas the one that controls all of these, and the laws, and cares for every aspect of things in the city, weaving everything together in the most correct way – this, embracing its capacity with the appellation belonging to the whole, we would, it seems, most appropriately call statecraft. (*Politicus* 305c–e, translated after Gill)

Statecraft thus turns out to be a kind of second-order expertise (Lane 1998, 141) that is distinct from the expertise of the lower-level crafts practised by the subordinate craftsmen. As Gill (2012, 195) puts it: 'The statesman directs the experts who are, as it were, the practical arms of his expertise ... In all these cases the statesman rules the other [subordinate] experts and they submit to him.' So, there is a kind of rule that can be described as "architectonic" where the subjects serve as the practical arm of the ruler in carrying out his aims. This is the statesman's manner of rule. Statesmanship rules over those subordinate crafts that engage in the mundane tasks of the city by coordinating their productive activities in the way that the conductor rules over the orchestra. The conductor doesn't play any of the instruments herself; rather, she regulates the behaviour of the other musicians so that what *they* play is a beautiful symphony. What the conductor's expertise consists in is not knowledge of how to play each of the instruments but in knowing what is appropriate in regard to the timing, tempo, and general "shape" of the piece.[10]

What I want to suggest is that this is also the manner in which Divine Reason 'rules over' Necessity in the production of the cosmos, not by directly involving itself in the practical tasks of generation, but by *managing* and *controlling* the causal interactions among subordinate material forces.

[10] Compare *Politicus* 283b–284b.

8.3 Reason's Rule in the Cosmos

As we have seen, the *Timaeus* holds that in the production of the world of becoming Necessity is 'employed in the service of the god as he attempts to bring to completion the character of what is best as far as it is possible' (46c). Reason is said to 'rule over' Necessity by 'persuading' it to follow a rational plan and that 'the result of Necessity's being overcome by wise persuasion was the initial formation of the universe' (47e–48a). According to Morrow (1950, 156), Reason's 'persuasion' consists in causing the simple bodies to act *in accordance with their own* necessary natures rather than forcing them to act in ways that are contrary to their nature:

> To say, then, that the cosmos comes about by persuasion means that it results from the working of the powers inherent in the materials of which it consists, each of them bringing into being the effects natural to itself, and none of them being under any constraint by a power outside nature. Nowhere in the series of events will we find any place where intelligence operates as a *deus ex machina* to subvert the natural sequence or replace the natural powers by powers or agencies of a different sort.[11]

But if all of the changes that cause things to come into being are the effects of necessary forces operating according to their own natures, what role is left for Reason in all of this? Here again Morrow's (1950, 157) insights are helpful:

> Intelligence is responsible, not for the specific effects of any given [necessary] power or set of powers, but for bringing together these specific powers into such juxtapositions and collocations that their joint natural effects are good, or at least better than would have been the isolated effects of the powers working in separation from one another. As our account of the carpenter shows, what the craftsman does is to bring together into the context of one event a variety of powers that would, in isolation, produce unrelated and irrelevant effects; but together they produce, with equal necessity, an object that expresses his purpose.

The idea, then, is not that Reason simply allows the simple bodies to behave in ways that express their own necessary natures. Instead he makes a difference to their motions by coordinating their necessary interactions thereby making them more orderly and structured so that together they bring about his good ends.[12]

[11] See also Johansen (2004, 99–103).

[12] Cf. *Timaeus* 69b2–c5. Johansen (2004, 100–1) also takes Reason to 'persuade' Necessity in the sense of '[c]hoosing, mixing, and coordinating the various necessary processes that are available to him given the nature of the simple bodies in accordance with his notion of what is best'. However, he

Let me fill this out a bit by returning to Plato's claim at *Timaeus* 47e that these necessary causes produce 'chance effects that lack order' when operating independently of intelligence (47e).[13] How can what is necessary also be associated with chance effects? To see what he has in mind, consider the famous market example that Aristotle uses in *Physics* II 6 to illustrate his theory of chance. In the example two people meet at the market and settle a debt. The creditor goes there to buy some fish, while the debtor goes there to see a show, and on the way they run into each other and settle their debt. Aristotle thinks that the correct way to analyse this situation is to think of it as composed of two independent causal processes (each man going to the market) converging in such a way that a separate result occurs (settling the debt). While each process has an intrinsic efficient cause (the creditor's decision to buy fish, the debtor's decision to see a show), Aristotle argues that the actual result does not and occurs entirely by accident. Of course, there is a scenario where the two men form a collective decision that allows them to coordinate their efforts in order to meet at that time and settle their debt. In that case the result does have an intrinsic efficient cause, their joint decision. But in the example the encounter came about entirely through the *accidental intersection* of two *independent* causal lines each with its own separate causes. It is in this sense that Aristotle says that chance is an incidental efficient cause of effects that appear as if they were operating for the sake of those ends (*Ph.* II 6, 198a1–13).

This analysis helps us see how necessity and chance are compatible with one another. Notice that the above analysis is compatible with the view that decisions necessitate behaviour so that each of the actors went to the market 'of necessity' (i.e. each man's being at the market was necessitated by antecedent states). The meeting would still be the result of chance, on Aristotle's account, since there was no separate decision that coordinated the two processes with a view to settling the debt. This explains how Plato can say that the simple bodies and their inherent powers can both operate according to necessity and that they produce 'chance effects that lack order' when undirected by intelligence. This also shows us how Reason can make

denies that the way Reason makes a difference to the necessary behaviours of the simple bodies is by making them more orderly and organised, since he denies that Necessity 'moves in an irregular and unpredictable way' before the Demiurge took over the cosmos.

13 With the following see Morrow (1950, 157–9) and Charles (2001, 258). I had formulated this part of my account, which draws a parallel with Aristotle's discussion of chance, before reading Morrow, whose own analysis helped me form the idea more clearly. I agree with Morrow that Plato does not explicitly provide an account of chance let alone one along Aristotelian lines. Nevertheless I think he is right to draw the connections he does, especially the point about the intersection of independent causal lines whose *internal* causal relations are governed by necessity.

a difference to the world while still allowing that all of the processes that cause things to come into being are the effects of necessary forces operating according to their own natures. The *Timaeus* appears to conceptualise the universe as a complex nexus of separate causal lines, each of which necessitates its outcome in a non-teleological way. What Plato is saying is that without Reason to coordinate them, those intersecting causal lines would never converge on a good outcome except by chance. And so, if necessity were the only force governing the universe, we should have expected it to be a chaotic and disorderly place. On this picture, then, Reason's 'wise persuasion' consists in actively coordinating separate causal streams, each of whose own internal causal relations are governed by necessity, so that they collectively produce a world that is orderly and good (something that would not have come about through necessity alone except by chance). Here Necessity is operating as a 'co-cause' and to say that Reason 'persuades' Necessity is just to say that it does all of this while working entirely within the limits of what is possible given the necessary natures of the materials involved.[14]

It is in this sense that Reason can be seen as playing the same architectonic role in generation that the statesman plays in governing the city. Like the statesman, Reason does not get its hands dirty with directly fashioning the universe. Instead the practical task of physically transforming the world is accomplished entirely by the simple bodies operating according to their own necessary natures. Reason's job is to regulate and control their necessary interactions by 'persuading' them to behave in such a way that what they produce is a world that is orderly, beautiful, and good.[15] Steven Strange (1999, 412–13) puts the point succinctly by saying that 'When Reason and Necessity combine [in the production of the universe], Reason "persuades" Necessity by producing the proper ordering of the mechanical causes to give the best results.'[16]

[14] Neither Plato nor Aristotle go beyond this sort of local determinism where the necessity involved applies only to the relation between cause and effect within each independent causal sequence (*A–B*, *C–D*). They do not embrace the more global causal determinism of the Stoics which applies to the relations between them. On the latter account necessity governs the entire network of causes. If that sort of global determinism were true, then Reason's 'persuasion' would either have no genuine effect on the world at all, since even the coordination of the separate causal streams would have been necessitated by prior causes, or else it would be seen as contravening necessity and forcing it towards his desired ends, in which case the metaphor of persuasion would be lost.

[15] Morrow attributes this architectonic function to the Demiurge while Mohr (1982, 41–2) assigns it to the World Soul. I deliberately speak in terms of "rational causes" or "Reason" to avoid this dispute. The point is the same in either case.

[16] One of the main pieces of evidence for linking the *Timaeus* with the *Politicus* is the fact that they both employ the conceptual distinction between 'primary cause' and 'co-cause' in their account of production, which seems to be peculiar to them. This distinction is absent from the *Phaedo*. At 99b2–3 Socrates contrasts 'cause' with 'that without which the cause would not be able to act as

In the final section of this chapter I shall argue that this is precisely how Aristotle himself envisions the causal relation between the formal and material nature of a biological substance during the process of its formation. My case for this is largely inferential. I argue, first, that Aristotle's account of generation, especially in the biological works, follows the *Timaeus*' model in several key ways. I then argue that, given these close connections, it is reasonable to infer that when Aristotle's speaks of formal natures 'using' necessity for the sake of an end he means to assign them the same architectonic role in substantial generation that Plato assigns to Reason in the construction of the universe.

8.4 Aristotle's Naturalisation of the *Timaeus* Model

Aristotle's teleological account of generation betrays its Platonic roots in at least three important ways: (1) in holding that an adequate theory of *genesis* must be able to explain how things come to be in a good and beautiful condition (which he also analyses in terms of order, arrangement, and the adjustment of means to ends); (2) in insisting that this requires appealing to some kind of order-producing cause; and (3) by downgrading the necessary forces of the Presocratics to the role of co-causes that are used by the order-producing cause. These connections with the *Timaeus* are familiar and so we can be brief. I start with the first two.

We saw above that the *Timaeus* places great emphasis on the fact that the cosmos is an ordered whole that exhibits beauty and goodness and that to account for this we must distinguish between 'those agents that possess understanding and thus fashion what is beautiful and good' (*kalôn kai agathôn*) from those 'that produce chance effects that lack order (*to tuchon atakton*)' when undirected by intelligence (*Timaeus* 46e). The contrast between beauty and goodness, on the one hand, and chance and lack of order, on the other, suggests that Plato is thinking of beauty and goodness here in terms of organisation and regularity. This fits well with the view expressed at *Gorgias* 503–506, where the good of each thing is associated with the proper ordering among its parts. Likewise, in *Metaphysics* I 3 Aristotle says that 'it is not likely that fire or earth or any other such being could cause things to be or come to be in a good and beautiful condition' (984b8–22). In that passage beauty and goodness are also associated with the order (*kosmos*) and arrangement (*taxis*) of things.

a cause'. But this is not the same distinction. Indeed, Socrates insists that to call the Presocratic causes 'causes' at all is absurd.

Now when Plato talks about the beauty and goodness of the cosmos, he has in mind (a) the orderly motions of the heavens, (b) the cycle of the seasons that follows upon them, and (c) living things and their parts whose generation and destruction are governed by those cycles. In *Laws* X he calls these 'the greatest and most beautiful things (*ta megista kai kallista*) in the world' (889a5–c5). And in the cosmological argument of the *Philebus*, he has Socrates argue that there must be 'a certain cause, of no small significance, that orders and coordinates the years, seasons, and months' and that this cause 'has every right to the title of wisdom and reason' (30c3–6). This is the thrust of Timaeus' claim that only those things that possess reason and intelligence are capable of fashioning 'what is beautiful and good' (*Timaeus* 47e).

Aristotle agrees with much of this analysis. His own use of *to kalon* in relation to the works of nature is explicitly connected with their teleological order. In this context, to say things come to be in a good and beautiful condition is to draw attention to the fact that living things and their parts exhibit the sort of adaptive complexity that promote their survival and well-being.[17] Presumably Aristotle also means to include under the beautiful and the good the orderly motions of the heavenly bodies and the regular cycle of the seasons that follow upon them. According to some scholars, Aristotle also thinks these things exhibit the same teleological adjustment of means to ends characteristic of living things. But I take him to mean that these other natural phenomena exhibit beauty and goodness only in the weaker sense of showing regularity and order (as opposed to being random and disorderly) without thinking that they also come to be for the sake of an end.

Aristotle also agrees with Plato that this calls for a special kind of efficient cause, one that is 'at the same time the cause of beauty and the sort of cause from which the change that belongs to things originates' (*Metaph.* 984b21–2). This is why they both praise Anaxagoras for saying there is *nous* in nature that is responsible for the arrangement of everything (*Phaedo* 97b5–98b5, *Metaph.* 984b15–22). Now Plato thinks that only rational causes are capable of generating this kind of order, since it requires looking towards eternal forms as its model (27d5–28b4). And since eternal being is the characteristic object of reason, it is only by possessing reason that something is capable of grasping

[17] *PA* I 1, 645a24–6. The argument of this chapter does not depend on whether we take Aristotle to be committed to the claim that there is beauty in the works of nature, as I do, or (as David Charles has suggested to me) that teleological order *plays the same role as* the beautiful in other domains. On the latter reading, Aristotle only means to suggest that the function of a thing (the end for the sake of which its parts come to be) is the source of value in biology just as *to kalon* is in other domains.

that sort of model.[18] This provides the basis for Plato's anti-materialism: the simple bodies and their inherent powers alone cannot account for the beauty and goodness of the visible world because they lack soul, and reason is only found in souls (*Timaeus* 46d4–8; cf. *Philebus* 30c3–6).

There is a sense in which Aristotle agrees with Plato that natural generation requires eternal forms as models. In *Metaphysics* VII 7 Aristotle says that the efficient cause of natural generation is the formal nature of the generating parent, which is 'specifically the same as the thing generated though it exists in another individual; for a human generates a human' (1032a24–5). He then argues in VII 8 that it is this feature of natural generation that makes Platonic Forms unnecessary (1033b26–1034a5). As I read that argument, what Aristotle is rejecting there is not so much the claim that generation requires a form to serve as a model – *Physics* 194b26–9 says that the form serves as a cause by being a *paradeigma* – but the claim that this form must be an independently existing entity that is separate from particulars. On Aristotle's theory the form that serves as the model does not exist in its own right apart from particulars; rather, it is the form that is embodied in the parent organism itself.[19] Thus there is no need to set up Platonic Forms as models because the formal nature of the particular generator is 'sufficient to produce and be the cause of the form that comes to be in the matter'. This form that serves as a model is the enmattered species form. And it serves as a model, not as some separately existing Form to which the nature of the developing organism looks in the manner of craftsmanship, but by being *the end towards which* it is internally directed.

More importantly, Aristotle rejects the Platonic view that intelligence is necessary for final causation. In *Physics* II 8, for example, he argues that nature acts for the sake of an end and then concludes the chapter by saying that it is absurd to deny that ends are present in nature on the grounds that we do not find any rational agents deliberating (199b26–30).[20] This reinforces the argument he made earlier that this is most obvious in the case of non-human animals (199a20–6). We observe them executing complicated

18 Johansen (2004, 72–4).

19 The thesis of *Metaphysics* VII 8 is that forms do not come to be (or pass out of existence) and so in this sense they are eternal.

20 Aristotle is clear in the *History of Animals* that when he speaks of animal intelligence (as he often does) he is talking about 'some other natural capacity' that is only *analogous* to human reason (e.g. *HA* VIII, 588a21–6). They are analogous, I suggest, in that both cause behaviours exhibiting the same teleological (means–end) structure (see Henry 2018, 16).

means–end behaviour without craft or inquiry or deliberation (*oute technê oute zêtêsanta oute bouleusamena*). For example, the orb spider builds her web for the sake of catching insects (she acts for an end), and yet she clearly does not reason out each step in the process with a view to that end. She executes those steps by nature, which is the internal efficient cause of her behaviour. On Aristotle's view, then, both reason and natures are capable of generating order, beauty, and goodness (cf. *Ph.* II 6, 198a3–4).[21]

The other key aspect of the *Timaeus* model that Aristotle appropriates is the distinction between primary causes and co-causes (see Chapter 4). In the biological works Aristotle regularly downgrades the simple bodies and their necessary interactions to the status of 'tools' that are used by the primary cause to bring about its ends.[22] We see this in Aristotle's criticism of Democritus' account of teeth-shedding in *GA* V 8. The reason Democritus goes wrong, he says, is that the necessary forces he invokes as causes of the change are causes only in the sense of proximate movers and instruments (*organa*) that are employed by the animal's nature for the sake of an end (789b3–15). In *DA* II 4 Aristotle makes the same point in the context of a claim about organic growth. Some say that fire is responsible for this change because it alone appears to feed itself and grow (416a10–12). Aristotle responds to this by explicitly invoking the conceptual machinery of the *Timaeus*: 'While fire is in a way a co-cause (*sunaition*) it is not the cause in the unqualified sense (*haplôs*), which is more properly the soul. For the growth of fire goes on without limit as long as there is fuel to burn, while in everything that is naturally constituted there is a limit and proportion to their size and growth. And this is due to soul, not fire, and to the form more than the matter' (416a13–18).

Aristotle's point here is that fire cannot be the primary cause of natural generation because its own natural behaviour is too unstructured and indefinite. As such, fire is incapable of producing the kind of order and complexity exhibited by natural generation on its own, and can do so only when operating as a co-cause under the direction of the animal's soul (or nature) which is the primary cause of the change.[23] As Plato puts it at

[21] In the case of elemental transformation and the meteorological cycles that follow upon them (e.g. rainfall), the efficient cause that is responsible for their order and arrangement is the annual rotation of the sun (*GC* 336a21–b15, *Meteor.* 346b20–347a6). Unlike the formal natures of living things, I have argued that this is not the kind of order-producing cause that operates for the sake of an end.

[22] In addition to the following see *GA* 740b25–741a3, where heating and cooling are explicitly compared to the craftsman's tools.

[23] Gill (commenting on the *DA* II 4 passage) writes: 'Soul [or nature] is a set of active capacities that directs the life activities, including nourishment and growth, for organisms. Fire contributes to the realisation of these ends as an instrumental (last) cause, but fire on its own cannot accomplish them

Timaeus 47e fire is among those inanimate agents that produce 'chance effects that lack order' when operating in the absence of a teleological cause. This, I argued, is the thrust of Aristotle's objection at *GC* 336a1–12. Because the materialists eliminate goal-directed efficient causes from their accounts they are forced to rely on the simple bodies whose generative powers are simply *too instrumental* to serve as primary causes of the change. Natural generation involves order, complexity, and the adjustment of means to ends, all of which are beyond the simple capacities of the elements themselves.

I want to extend the argument here by suggesting that Aristotle adapted the *Timaeus* model in a further way, namely, by assigning the formal natures of living things the same architectonic role in substantial generation that Plato assigned to Reason in the creation of the whole world-order. Recall that according to the *Timaeus* Reason exerts its influence on the world of becoming, not by directly moving the materials itself, but by 'persuading' those materials to direct their own necessary capacities towards what is best. There are at least two passages in Aristotle's biological works that are strongly reminiscent of this idea. Consider first *PA* 663b22–4 (cf. b28–35): 'Let us now speak about how the necessary nature, whose products occur from necessity, has been put to use by the formal nature for the sake of something.' This is Aristotle's version of the wise persuasion principle from *Timaeus* 47e3–48a7. Of course, he does not use the language of nature 'ruling over' necessity (though see *GA* 770b15–17) and 'persuading' it to act for an end. Instead he uses the more mundane expression *katachraomai* (nature 'puts [necessity] to use'). But that's what we should expect if Aristotle is attempting to naturalise the *Timaeus* model. However, it is interesting that what uses necessity here (what I have translated as 'the formal nature') is *hê kata ton logon phusis*, literally 'the rational nature' or, to use a less loaded translation, 'the nature according to the account'. Although Aristotle does not mean to imply that natures literally possesses reason – natures, Aristotle assures us, do not deliberate about their ends (*Ph.* 199b26–30) – this expression would have easily called to mind the *Timaeus*.[24]

because fire grows without bounds as long as there is material to burn. The natural behaviour of fire *must be constrained and directed* [sc. by soul] so that its natural behaviour subserves a particular end. Without direction the unbridled action of fire will not yield the appropriate outcome. Soul plays the supervisory role, and in this sense soul is the first efficient cause of nourishment and growth' (Gill 1989, 200). This is the essence of the architectonic model of substantial generation. While Gill doesn't make the connections to the *Timaeus* or to Plato's political philosophy that I do, her own interpretation of the *Politicus* certainly invites them (Gill 2012, chapter 6).

24 Leunissen (2010, e.g. 202, 209) downgrades this pattern of teleological causation to what she calls 'secondary teleology'. I discuss her interpretation in the Appendix to this chapter.

My other key text is *GA* II 6, 743a36–b18. This passage is at the very core of Aristotle's teleological model of substantial generation.[25] Aristotle has just finished explaining how the formation of the uniform parts is effected by the agency of heating and cooling. The different ways these forces interact with the materials involved explains why the parts have features like hardness and softness, elasticity, brittleness, and so forth. Towards the end of his account Aristotle qualifies those explanations by saying that reference to heating and cooling alone cannot explain the formation of the parts because they do not account for the timing, location, and degree of those heatings and coolings or why certain kinds of materials are made available to them: 'The heating that makes flesh or bone, however, does not operate on any chance matter in any chance location at any chance time but only those that are suited by nature.'[26] This is followed by the claim that the right degree of heating and cooling to suit the parts is supplied by the nature of the generating parent just as the cook regulates the amount of heating and cooling to suit the dish she's preparing (743a26–36; cf. 767a13–28). Aristotle then generalises the point at 743a36–b18, where he attempts to locate the accounts of the parts in terms of so-called material necessity within a broader teleological framework:

> Now cold is a privation of heat. And both of these are used by nature insofar as they have a capacity for making one thing this way and another that way *of necessity*. So that in the context of development, at any rate, it is *for the sake of something* that the heating and cooling of these and the formation of each of the parts takes place, the flesh being made soft – heating and cooling making it such, on the one hand, from necessity and, on the other hand, for the sake of something – while sinew is made solid and elastic and bone solid and brittle[27] . . . We must state then, as we have said, that all of these things come into being, on the one hand, by necessity, and on the other hand, not

[25] For an earlier version of the ideas presented here see Henry (2007b, 198–9). For a similar reading see Gotthelf (2012, chapter 4, especially 104–10). Gotthelf's treatment of that passage originated in part from comments I made on an earlier draft of his paper in which I urged him to consider that the principle articulated at *GA* 743a36–b18 held the key to his worry that *GA* II 6 might provide evidence for the view (defended by Charles 1988) that Aristotle thinks a 'full material-level account' of morphogenesis might be possible in principle. My suggestion to Gotthelf was roughly the account that follows (as he generously acknowledges: Gotthelf 2012, 90, n. 105, n. 27).

[26] *GA* 743a21–36.

[27] Compare *Politicus* 305c–e (discussed above): 'For true kingship must not perform practical tasks itself but control those with the capacity to perform them because it knows when it is the right time to begin and set in motion the most important things in cities and when it is the wrong time. And the others must do what has been prescribed for them.' Aristotle has just said that heating and cooling alone are insufficient to account for morphogenesis because they cannot account for things like appropriate timing or due measure.

by [that kind of] necessity but for the sake of something.[28] (*GA* II 6, 743a36–b5, b16–18)

Aristotle tells us here that (i) heating and cooling have the capacity to necessitate (in a non-teleological way) certain kinds of changes but that (ii) in the context of animal generation these necessary capacities are used by the animal's formal nature for the sake of bringing about certain ends. All of this echoes the point of *DA* 416a13–18 that elemental forces are only capable of producing the kind of order and complexity exhibited by natural generation when operating as *co-causes* under the direction of the animal's formal nature, which is the primary cause of that change. Together this points to the idea that the *Generation of Animals* is working with the same architectonic model of generation that Plato develops in the *Timaeus*.

Aristotle's account of how membranes form around the embryo early on in generation helps to illustrate this principle behind *GA* 743a36–b18. At *GA* 739b26–32 he tells us that the membranes occur 'both from necessity and for the sake of something'. On the one hand, the fluid nature of the embryo is such that its surface *must* solidify when exposed to the right degree of heating and cooling resulting in a thin membrane being formed around it. The necessity involved here is material necessity: the nature of the causal powers involved are such that their mutual interaction *necessitates* that effect in a non-teleological way (see *GA* 778b16–18).[29] On the other hand, experience shows that the early embryo must be separated from its fluid environment if it is going to survive to maturity; it cannot survive otherwise. So the presence of extraembryonic membranes at that particular stage of development is also necessary for the sake of an end (what Aristotle calls *conditional* or *hypothetical* necessity). The principle articulated at *GA* 743a36–b18 shows us how to integrate these two parts of the account into a unified explanation in a way that follows the "twin birth" explanations of the *Timaeus* (47e3–48a7, 68e1–69b1; Aristotle adapts this pattern of explanation for biology at *PA* 642a1–b5).[30] Heating and

28 Again compare the dual pattern of explanation recommended at *Timaeus* 47e3–48a7 and 68e1–69b2.

29 As mentioned in connection with the *Timaeus*, by treating material necessity as a co-cause in the realisation of some end I do not mean to suggest that it can be 'subsumed under' conditional necessity (cf. Leunissen 2010, 24, 99). Even in the context of teleological explanations it still operates as a form of non-teleological necessity, though one that is capable of being 'persuaded' to work for an end (to borrow a phrase from the *Timaeus*). When heating and cooling act on an earthy fluid, the mutual activation of the causal powers involved necessarily produces a membrane. This necessary causal relation is not conditional on an end; those powers will still produce that effect whether or not the capacity to necessitate that effect is being used to achieve some goal.

30 For further discussion of the *PA* passage see the Appendix to this chapter.

cooling have the capacity to necessitate a membrane around the embryo, which they have independently of any reference to natural goals. But in order to achieve the goal of separating the embryo from its fluid environment those capacities must be brought to bear on the process at the right time, in the right location, and to the right degree. Aristotle thinks that these features of the change are not due to material necessity alone but depend on the goal-directed activity of the animal's formal nature. While heating and cooling might be responsible for physically transforming the embryo by making it soft or hard, brittle or elastic, etc., developing embryos would not consistently be exposed to the right degree of heating and cooling, at the right location, and at the right time unless the entire process was being regulated and controlled by the formal nature operating according to conditional necessity (cf. *GA* 734b34–735a4). That sort of coordination, Aristotle thinks, cannot be achieved by the simple bodies operating according to their own necessary natures with the sort of consistency exhibited by natural generation but only when they function as co-causes that are used by nature for the sake of an end.

In this section I have argued that both Plato and Aristotle defended an architectonic model of generation, which I claimed has its roots in the politics of the *Politicus*. Both sought to explain the generation of natural substances in terms of the causal interactions between the simple bodies operating according to material necessity but in a context where they are put to use by some other order-producing cause (Reason for Plato, the natures of living things themselves for Aristotle). On this model all of the "hands-on" labour of building a substance is assigned to various physical forces whose necessary causal interactions physically transform the matter in various ways, while the choreography of those interactions – their timing, location, order, and so forth (what Plato in the *Politicus* calls their 'due measure') – is due to the goal-directed actions of that other principle that uses those forces as instruments and co-causes for the sake of realising its good ends.

8.5 The Role of Material Natures in Non-Teleological Contexts

To this point I have been concerned mainly with those aspects of biological generation that Aristotle thinks must be explained by reference to final causes (so-called adaptations), e.g. eyes, hands, wings, and so forth. In this context Aristotle thinks the formal nature of the organism plays a regulatory ("architectonic") role in development by coordinating the necessary motions and changes of the simple bodies (e.g. heatings and coolings) so that what *they* bring about is a functionally organised substance whose parts contribute to its survival and well-being. However, elsewhere I have argued that it is not the case that Aristotle thinks all natural, let alone all biological, phenomena are determined by final causes. And so not all explanations that invoke the simple bodies and their inherent powers will have to be embedded within a broader teleological framework. Instead, some features of living things can be explained by reference to properties associated with their material nature alone without having to appeal to other facts about their formal nature. In these cases, Aristotle even appears to treat the material nature as causally basic in the sense of being a cause of many other features of a living thing while nothing more fundamental is the cause of it.[31] The upshot of this is that both formal *and* material natures can serve as first principles of scientific explanations. I have discussed this in detail elsewhere. I briefly mention two examples from the biology that help illustrate this point.[32]

8.5.1 The Pathêmata *of* GA *V*

In *GA* I–III Aristotle examines the development of the various parts of animals, both those that belong to animals in common (e.g. heart, liver) and those that belong separately to particular kinds of animals (e.g. teeth, hair, skin). In Book V he turns his attention from the formation of the parts themselves to the causes of their various 'attributes' (*pathêmata*), specifically, those attributes that distinguish the parts of animals from one another by the-more-and-less (e.g. differences in the *colour* of the eye, in the *length* of the hair, in the *texture* of the skin).[33] In these cases, Aristotle says, we must have recourse to a different pattern of causal explanation:

> About these [attributes] and all the things of this sort, one must no longer suppose that the same manner of explanation is involved. For all those which are neither the common products of animal natures nor are proper to each kind – of those sorts of things, none either is or comes to be for the sake

[31] See Henry (2008). This position is also defended by Charles (2001, e.g. 334–5), Leunissen (2010, e.g. 24–5, 2012a), Gotthelf (2012, chapter 9), Connell (2016, chapter 10). By contrast, Scharle (2008, 30–5) argues that the material nature of a thing is always causally dependent on its formal nature and that material-efficient causation always requires formal-final causation. On this view no natural phenomenon can be fully explained without recourse to final causes; explanations in terms of material natures must always "bottom out" in teleological explanations.

[32] In addition to these two, see Aristotle's discussion of egg-layers in *GA* III. I discuss that example in Henry (2008).

[33] For a discussion of the more-and-less in Aristotle's biology see Lennox (2001a, chapter 7).

> of something. For while an eye is present for the sake of something, its being blue is not (unless this attribute is a proper feature [*idion*] of the kind).[34] In some cases [the attribute] does not reach to the definition of the substantial being but the explanations must be referred back to the matter and the source of motion as coming to be from necessity. (778a29–b1, Leunissen and Gotthelf translation with modifications)

Aristotle is explicit in this passage that not all features of animals are present for the sake of something, and so not all attributes are subject to teleological explanation. In some cases the cause of the attribute 'must be referred back to the matter and the source of motion as coming to be from necessity'. The type of necessity Aristotle has in mind here is so-called material (or Democritean) necessity (cf. *Ph.* 198b11–14), which he goes on to distinguish from the (logical) necessity involved in his essentialism: 'Thus, animals have eyes from necessity for having that sort of part is present in the substantial being of an animal, while they have a particular sort of eye also from necessity though not the same kind of necessity; rather, it is owing to another mode of necessity, namely, because it is naturally suited to act and be acted upon in this or that way' (*GA* 778b16–18; cf. 782a20–3). As we have seen, the use of material necessity as a co-cause of generation is a central feature of the architectonic model developed above. The idea that certain *pathêmata* might be caused by material necessity alone was already suggested back in *GA* II 1 where Aristotle said that heating and cooling can produce the *pathê* of the parts (e.g. their hardness and softness) but that on their own they could not generate the form (*logos*) that makes each part what it is.[35]

It is tempting to dismiss *GA* V as a mere catalogue of accidental variations. But the attributes under investigation not only include sub-specific differences (e.g. variations in human eye colour) but also features that belong in common to whole kinds (778a20–1). In the latter cases

[34] By *idion* here Aristotle does not simply mean a feature that belongs in common to all members of a species but what later Aristotelians called a *proprium*: a property that is not part of a thing's essence but 'belongs to it alone and is convertible with it' (*Topics* I 5, 102a18–20). Such features serve to distinguish the kind from everything else, even if they do not define it. The capacity to learn grammar is an *idion* feature of humans in this sense (102a20–3), as is the capacity to ink in cephalopods (*PA* 679a4), the ability of snakes to turn their heads 180 degrees without moving the rest of their body (*PA* 691b31), and feathers in birds (*PA* 692b10). By contrast, having black eyes is not an *idion* feature of cows. For, although all cows have black eyes, having black eyes is not 'proper' to cows in the technical sense for the feature extends wider than the kind and so the two terms are not convertible. For an alternative way of reading this qualification see Leunissen and Gotthelf (2010, 334 n. 15). If my reading is correct, then (pace Leunissen and Gotthelf) being common to all members of a species is not even a necessary condition for teleological causation.

[35] Cf. *GA* II 4, 734b17–735a4. For a discussion of this passage see Gotthelf (2012, 95ff.).

Aristotle appears to be after genuine causal explanations that apply at the level of the widest kinds to which those features belong, which suggests that the features in question have intrinsic causes (cf. *APo.* I 5). In *GA* V 1, for example, Aristotle seeks to 'grasp the universal (*kathalou*) concerning the differences between eyes, on account of what cause some eyes are blue, some light-blue, some goat-eyed, and some dark-eyed' (779b12–15). Aristotle is not (or not only) talking about differences within a single species but about variations in eye colour among all animals that possess eyes (the wider kind). Why don't all animals have the same colour of eyes? Why do cows have dark eyes and sheep grey eyes, while horse eyes are sometimes heterochromatic? Again, *GA* V 3 investigates the causes on account of which hedgehogs (as a kind) have long, hard spines while rabbits have short, fine hairs. Finally, in *GA* V 8 Aristotle inquires into the causes of why some kinds of animal (e.g. humans) shed their first set of teeth and then goes on to offer a teleological account of the phenomenon. So, while *GA* V certainly does contain accounts of individual material accidents (e.g. grey hair), many of the attributes under investigation are kind-specific features that are present owing to intrinsic causes.[36]

Aristotle's account of curly fur (e.g. sheep's wool) provides a good illustration of the pattern of explanation typical of *GA* V. The curl is said to result from the interaction between three causal factors: (1) the natural inclination of the earthy element in the hair to go down; (2) the natural inclination of its fiery element to go up; and (3) the weakness of the hair itself. When hair is subject to the contrary impulses of its elements (the 'double-motion'), it twists under the pressure owing to its weakness resulting in a curl (*GA* V 3, 782b19–24). There is no mention of any final causes at work here nor is there any suggestion that a teleological explanation is forthcoming. In this case reference to the necessary natures of the materials involved is taken as sufficient to explain why the hair curls. Of course, one might respond by insisting that the account as presented in the text is simply incomplete and that, if pressed, a teleological explanation could be given. But this response would beg the question. For to call the account incomplete is to assume that Aristotle thinks all explanations that invoke material natures must ultimately bottom out in final causes, which is precisely the point in dispute. Given Aristotle's methodological remarks at the beginning of *GA* V 1 (778a29–b18), our default assumption should be that the features discussed in *GA* V are *not* present for the sake of

36 Leunissen and Gotthelf (2010) provide an extended defence of this point. See also Pellegrin (1986, 157).

something unless there are strong reasons for thinking otherwise (e.g. the ears of seals in V 2 and shedding teeth in V 8).

8.5.2 The Omentum

The other example I want to mention where the material nature of an animal makes a positive contribution to its development independently of the teleological actions of its formal nature is the omentum, which is a functional membrane that plays a role in digestion. This example is interesting because, while the omentum contributes to the animal's overall functioning, there is no sense in which this part can be said to have *come to be* for the sake of that end. Its presence in the animal is the result of material necessity alone:

> This part comes into being from necessity (*ex anankês*) in the following way. When a mixture of dry and moist is exposed to heating, the surface becomes skin-like and membranous, and this location is full of such nutriment. Further, on account of the thickness of the membrane, it is necessary for the filtrate of the bloody nourishment to be fatty (for this is thinnest) and to become hard or soft fat rather than a fleshy and bloody composition when it undergoes concoction owing to the heat in this region. The coming-to-be of the omentum, then, happens in accordance with this account. But nature uses it (*katachrêtai*) for proper concoction so that the animals may concoct their nourishment easily and quickly. For what is hot has the ability to effect concoction, and what is fatty is hot, and the omentum is fatty. (*PA* IV 3, 677b21–33)

Although there is a reference to final causes, it is unrelated to the process of formation. Aristotle is clear here that the omentum *comes to be* as a result of material necessity alone. Once it has come to be, the nature (nutritive soul) of the mature animal *uses* the omentum for the sake of digestion. But this acquired function is not part of the explanation for why that part comes to be present in the first place.

Contrast this with Aristotle's account of why membranes form around the embryo during development at *GA* 739b26–32 (see above). Those membranes result from the very same chemical reaction that produces the omentum. The difference is that in that passage Aristotle is explicit that those membranes come to be *both* of necessity *and* for the sake of something. The fluid nature of an embryo is such that its surface *must* solidify when exposed to the right degree of heating and cooling generating a thin membrane around it. The developing embryo's formal nature uses this necessity in order to realise the goal of separating itself from its fluid environment. In this case the function those membranes perform in development explains why they come to be in the first place. The omentum passage makes no such connection: Aristotle says only that the omentum is produced from necessity. It is true that this part is later *used* by the animal's nature (nutritive soul) for the sake of digestion. But – and this is the crucial point – that acquired function plays no role in the explanation for why that part comes to be or why it has the features that it does. Its presence in the animal is completely determined by the antecedent material forces that necessitate its existence. To use an analogy, the pond behind our house freezes every winter owing to the fact that when water drops below zero degrees Celsius it necessarily freezes into ice. Once it is frozen we can use the ice for the sake of playing hockey. But that acquired function plays no role in explaining why the ice was formed in the first place (it was not the goal for the sake of which it came to be).[37]

So it turns out that Aristotle does not think all explanations that invoke material natures must be embedded within a broader teleological framework. There are some features of living things that can be explained by reference to properties associated with their material nature alone without having to appeal to other facts about their formal nature. This is not something the received view of Aristotelian teleology typically recognises.

Appendix to Chapter 8

In this brief Appendix I want to discuss how the interpretation of Aristotle's natural teleology defended in this book (especially Chapter 8) compares with another interpretation that is currently in fashion.

One of the most influential accounts of Aristotle's teleology in the recent literature is the reading developed by Mariska Leunissen (2010). According to Leunissen, Aristotle distinguishes between two distinct patterns of teleological causation. Both patterns involve 'the goal-directed actions of [the formal] nature as an internal efficient cause'. But in the one case this nature acts by 'realizing an internal, preexisting potential for form' (what

[37] Compare Johnson's (2005, 150–8) analysis of the rainfall example. Leunissen (2010, 95, 140–1) treats the omentum as a case of what she calls 'secondary teleology', where the formal nature uses *raw materials* that are already present of necessity for the sake of some end during the process of its formation. This is misleading since here the goal-directed activity of the animal's formal nature plays no role in the formation of the omentum either in producing the raw materials for constructing that part or in the actual formation of the part itself. Final causes are simply not involved in the account of *the coming-to-be* of the omentum. I discuss Leunissen's influential account of Aristotle's natural teleology in the Appendix to this chapter.

she calls primary teleology). In the other it acts 'by using materials produced by material necessity for the good of the natural being' (what she calls secondary teleology).[38] The first pattern applies to those features of an organism that are essential for being the kind of substance it is (those that are present 'for the sake of exercising their vital and essential capacities' or 'for contributing to the performance of those vital and essential capacities'). The second is active in the production of those features that enhance the organism's well-being but are not technically required for its existence (what Leunissen calls 'subsidiary and luxury features'). The categories of primary and secondary teleology are thus divided along two axes:

(A) *According to the type of causation involved in the part's formation.* Primary teleology involves the actualisation of a pre-existing potential for form involving conditional necessity, while secondary teleology involves 'using' resources that are present owing to material necessity alone.

(B) *According to the type of part that results.* Primary teleology underwrites the formation of vital and essential parts, while secondary teleology underwrites the formation of subsidiary and luxury parts.

This matrix of distinctions forms the heart of Leunissen's interpretation of Aristotle's natural teleology.

The interpretation defended in this book agrees with the basic intuition behind Leunissen's account, namely that to understand Aristotle's natural teleology we should focus on the different patterns of causal interactions that occur between the goal-directed activities of an organism's formal nature and the necessary changes arising from its material nature. What I am sceptical about is the claim that Aristotle himself means to collect these differences into two discrete and mutually exclusive categories (especially the way Leunissen collects them). In what follows I will focus mainly on Leunissen's account of the teleological causation involved in biological generation. I don't take issue with the claim that Aristotle makes a distinction between 'vital and essential' parts, on the one hand, and 'subsidiary and luxury' parts, on the other. Rather, what I deny is that the generation of these two kinds of parts is due to two *distinct kinds* of teleological causation. Instead, I shall argue that her categories of primary

[38] Leunissen (2010, 208–9). When Leunissen talks of the formal nature 'realizing an internal, preexisting potential for form' this is meant to be analogous to a craftsman following a blueprint that specifies the form of the product. When she goes on to associate the products of primary teleology with 'vital and essential parts', she means that these parts are specified in advance by that blueprint (they are included in its essence or substantial being).

and secondary teleology (suitably circumscribed) are simply variations on the architectonic model of generation presented in Chapter 8.[39]

The main problem I find with Leunissen's two patterns of teleological causation is that they obscure important similarities and differences between the cases she examines. Consider the following two examples, both of which are treated as products of secondary teleology:

> Case 1 *Horns, teeth, and tusks.* What is bony in the bodies of animals is in origins earthen; this is also why it is not abundant in the largest animals, to speak with a view to what occurs for the most part. For the residual surplus of this sort of body, being present in the larger of the animals, is used by nature for protection and advantage, and the surplus, which flows of necessity to the upper region, in some cases it distributes to teeth and tusks, in other cases to horns. (*PA* III 2, 663b28–35, Lennox translation)

> Case 2 *Omentum.* This part comes into being from necessity (*ex anankês*) in the following way. When a mixture of dry and moist is exposed to heating, the surface becomes skin-like and membranous [this membrane is the omentum itself], and this location is full of such nutriment. Further, on account of the thickness of the membrane, it is necessary for the filtrate of the bloody nourishment to be fatty (for this is thinnest) and to become hard or soft fat rather than a fleshy and bloody composition when it undergoes concoction owing to the heat in this region. The formation of the omentum, then, happens in accordance with this account. But nature uses it (*katachrêtai*) for proper concoction so that the animals may concoct their nourishment easily and quickly. For what is hot has the ability to effect concoction, and what is fatty is hot, and the omentum is fatty. (*PA* IV 3, 677b21–33)

I discussed Case 2 at the end of Chapter 8. In what follows, I will assume for the sake of argument that Leunissen's general interpretation of Case 1 is correct and that in these cases the formal nature makes use of leftover resources that are present owing to material necessity alone.

Leunissen finds in these two cases a common pattern of teleological causation where some materially necessitated outcome *X* (an outcome that comes to be by material necessity alone) gets co-opted by the animal's nature for the sake of an end *E*.[40] If we focus simply on this abstract pattern, then the two cases do appear quite similar to one another. In Case

[39] The following expands on my review of Leunissen's book in Henry (2011). These ideas first arose from our correspondences on early drafts of her manuscript.

[40] Leunissen also treats these as 'subsidiary and luxury parts' that are not specified in advance by the blueprint of the animal (see note 38 above). Lennox (2001c, 249) disputes this claim. He thinks that the reference to 'nature according to the account' just before this passage (663b20–4) strongly suggests that he is thinking of these parts as included in the formal account of the animal. If this is right, it calls into question the (B) part of the matrix above.

1, the formal nature of the animal uses certain raw materials that are present owing to material necessity in order to make parts suitable for protection and advantage. In Case 2, the membrane that results from material necessity gets used by the mature organism for the sake of enhancing digestion. But this depends on comparing the two cases from two different perspectives. In the first case we are talking about the generation of the part (where *X* is the raw materials used to make the part and *E* is the part itself) whereas in the other case we are talking about how the part functions in the mature organism after it has been generated (where *X* is the part itself and *E* is the work it does). When we compare the two cases from the same perspective of the part's formation, however, we can see that the causation involved is radically different.

To see this, consider Aristotle's account of the uniform parts in *GA* II 6 which culminates in the following passage:

> Case 3 *Flesh, sinew, bone.* Now cold is a privation of heat. And both of these are used by nature insofar as they have a capacity for making one thing this way and another that way *of necessity*. So that in the context of development, at any rate, it is *for the sake of something* that the heating and cooling of these and the formation of each of the parts takes place, the flesh being made soft – heating and cooling making it such, on the one hand, from necessity and, on the other hand, for the sake of something – while sinew is made solid and elastic and bone solid and brittle . . . We must state then, as we have said, that all of these things come into being, on the one hand, by necessity, and on the other hand, not by [that kind of] necessity but for the sake of something. (*GA* 743a36–b5, b16–18)

Leunissen agrees that this pattern of teleological causation should be associated with her category of primary teleology.[41] Here, while the generation of these parts still involves the necessary interactions of the simple bodies and their inherent powers, the formal nature uses that necessity as a tool in realising those ends. This does not strike me as significantly different from Case 1 above. In both cases the part in question *is fashioned* by the goal-directed actions of the animal's formal nature, which adapts its shape and structure to the function it is meant to perform. And both involve using material necessity for the sake of that end. By contrast, Aristotle is clear in *PA* IV 3 that the *genesis* of the omentum is explained entirely in terms of material necessity; the goal-directed actions of the formal nature are not responsible for fashioning that part. While the omentum is *used* by the mature animal for the sake of digestion, it did

41 See Leunissen (2010, 88 n. 22).

not come to be for that reason; that acquired function plays no role (as aim) in the process of its formation.[42]

From the perspective of generation, then, the relevant distinction seems to be between those cases where the formal nature is the primary efficient cause and material necessity serves as a co-cause (Cases 1 and 3) and those cases where the part comes to be through material necessity alone (Case 2). Grouping the omentum together with horns, teeth, and spurs as products of so-called secondary teleology effaces this important distinction.

Leunissen's characterisation of secondary teleology is problematic in another way, too. For example, she argues that the products of secondary teleology 'can only be explained satisfactorily by reference to the goal-directed actions of nature as an internal efficient cause' (Leunissen 2010, 25). But this is not true for parts like the omentum. As we have just seen, the only efficient causes at work in its formation are the necessary interactions between the simple bodies and their inherent powers (those associated with the animal's material nature).[43] More importantly, she insists that the products of secondary teleology '*come to be* of material necessity but are *present* for a function' (Leunissen 2010, 95, 201). While this is true for the omentum, it does not seem to be true for horns, teeth, and spurs (Case 1). Even if we agree that the animal's formal nature makes these parts using resources that are lying around thanks to material necessity, why should that fact entail that the resulting parts do not *come to be* for the sake of protection and advantage? Suppose that as a necessary consequence of how lumber is packaged a builder discovers that she has leftover materials that were not specified in the original blueprints. If she uses that extra lumber to build window boxes, we would still say that those window boxes *came to be* for the sake of holding flowers (that she made the boxes *in order to* hold flowers). Likewise, we can still say that the animal's nature makes horns for the sake of protection even if it is true that it makes them out of surplus materials that are present owing to material necessity alone.

Leunissen argues that what unifies her two categories is the fact that the parts in question are all products of the goal-directed actions of the

[42] I would argue that menstrual blood, which is another star example of Leunissen's secondary teleology, is closer to the omentum in this respect (see Leunissen 2010, 12, 91). One might try to make a finer-grained distinction here between Case 3, where the formal nature actively coordinates the necessary interactions of the simple bodies and their inherent powers in order to realise an anticipated goal, and Case 1, where nature repurposes some materially necessitated resources for an unanticipated goal (which is how Leunissen reads *PA* 663b20–35). Even if Aristotle meant to draw that fine-grained distinction, it does not strike me as significant enough for treating these as mutually exclusive kinds of teleological causation. See below.

[43] Leunissen (2010, 140): 'The formal nature of the animal plays no role in this part's coming-to-be.'

animal's formal nature. Leaving aside cases like the omentum, the whole distinction between her two kinds of teleological causation seems to come down to the fact that in some cases the formal nature makes the materials it needs to form the part (primary teleology) while in other cases it makes use of materials that are already present owing to material necessity (secondary teleology).[44] I agree that what is interesting about the latter examples (e.g. horns) is that they bring into focus how formal natures can be opportunistic in exploiting resources that happen to be lying around. But that does not strike me as a significant enough feature of those cases to justifying classifying them as a *distinct kind* of teleological causation. Instead, I see Leunissen's two categories simply as variations on what I have called the "architectonic" model of generation, where the formal nature uses material necessity as a co-cause in bringing about its good ends – a model that I have argued has its roots in Plato's *Timaeus*. This is distinguished from those other cases (e.g. Case 2) where the formation of the part is due entirely to the necessary interactions of the simple bodies and their inherent powers alone without any influence being exerted on the process by the goal-directed actions of the formal nature.

This difference between our two interpretations of Aristotle's natural teleology can be seen by comparing the principles expressed in *GA* 743a36–b18 (translated above) and *PA* 663b20–4. The latter reads: 'So, then, what the nature of the horns is for the sake of has been stated, and owing to what causes some have such things while others do not; but since there is a necessary nature as well, let us now say how the necessary nature, whose products occur from necessity, has been put to use by the formal nature for the sake of something.' Leunissen associates *GA* 743a36–b18 with primary teleology, while she takes *PA* 663b20–4 as a description of secondary teleology.[45] By contrast, I take the two passages to express the same basic teleological axiom, which I connected with the architectonic model of generation in Chapter 8. There are two reasons for taking them in this way.

[44] Leunissen (2010, 88–9, 202): 'Aristotle describes the action of the formal nature of the animal [in secondary teleology] as *using* materials that are present of necessity for a good purpose, rather than as *producing* those materials for the sake of the realization of some function [as in primary teleology].' She builds this distinction from her elaborate interpretation of the analogy with housekeeping at *GA* II 6, 744b11–27. However, as she is careful to point out, Aristotle himself is not concerned with distinguishing kinds of teleological causation in that passage. Her reading also seems to have been strongly influenced by Lennox's (2001c) comments on *PA* 663b20–4.

[45] Leunissen (2010, e.g. 202) takes the reference to necessity in the disputed passage at *PA* 663b20–4 to indicate that the surplus materials being used to make the parts in question are present owing to material necessity alone. She associates *GA* 743a36–b18 with primary teleology at 2010, 88 n. 22.

First, the pattern of teleological causation described in *GA* 743a36–b18 is said to apply to 'the formation of *each* of the parts' (*ginesthai tôn moriôn hekaston*). Even if we confine this remark to the uniform parts discussed just before this passage, this still includes nails, horns, hoofs, and bills (743a14–16), which Leunissen treats as products of secondary teleology. But it also includes vital parts like flesh, bone, and sinew, which she associates with primary teleology. So Aristotle appears to think that the formation of all of these parts is underwritten by the same form of teleological causation where the formal nature of the developing embryo uses material necessity for the sake of something.

Second, *PA* 663b20–4 strikes me not as a reference to some subsidiary form of teleology but as a statement of Aristotle's commitment to the method of inquiry recommended at the end of *PA* I 1, which he says we ought to follow if we want to 'grasp the causes' (*labein tas aitias*) (642b2–4). As we shall see, this is the very same method found at the end of *GA* 743a36–b18. This supports my claim that these two passages in fact describe the same basic teleological axiom.

At the end of *PA* I 1 Aristotle tells us that 'there are these two causes, the final cause and the necessary cause' (642a1–2) and that when explaining the parts of living things 'one should *first and foremost* succeed in stating both of these' (a14–15). The example of breathing that Aristotle uses to illustrate the point makes it clear that the necessity in question is material (rather than conditional) necessity. Aristotle closes *PA* I 1 by saying that this mode of inquiry (*tropos houtos ho tês methodou*), where some feature of the organism is said to come about *both* of necessity *and* for the sake of something, is the one we ought to follow if we want to 'grasp the causes' (*labein tas aitias*) (642b2–4).

The *PA* I 1 method is nicely illustrated by Aristotle's account of extra-embryonic membranes from *GA* II 4:

> Case 4. *Extraembryonic membranes.* When [the bodily substance in the menstrual fluid is 'set' by the heat in the semen], as the earthy constituent solidifies membranes form all around its outer surface. This occurs both from necessity and for the sake of something. For [1] it is necessary that when the surface of a thing is exposed to heating and cooling it necessarily (*anankaion*) solidifies. And [2] there is a need (*dei*) for the developing embryo not to be in liquid but to be separated from it. (*GA* II 4, 739b26–32)

In [1] Aristotle invokes the same necessary law of chemistry at work in the formation of the omentum (Case 2), which says when a mixture of dry and moist is exposed to heating and cooling the surface necessarily becomes skin-like and membranous. However, in this case this necessary process is used by the formal nature to satisfy an existing biological need. Aristotle does not restrict this dual pattern of explanation to subsidiary or luxury parts but associates it with a feature of the embryo that is conditionally necessary for its survival.[46]

What is important for our purposes is that this same norm of inquiry is reflected in both of the disputed passages at *GA* 743a36–b18 and *PA* 663b20–4. First consider the discussion of horns in *PA* III 2. The overall structure of the chapter again illustrates how inquiry in biology should 'first and foremost' be conducted with an eye to discovering both necessary and final causes. In the first part of the account (662b23–663a19) Aristotle explains facts about horns by reference to their function, which identifies their final cause. He then introduces the second part of the account with the disputed remark: 'So, then, what the nature of the horns is for the sake of has been stated, and owing to what causes some have such things while others do not; but since there is a necessary nature as well, let us now say how the necessary nature, whose products occur from necessity, has been put to use by the formal nature for the sake of something.' This is a clear statement of Aristotle's commitment to the method of inquiry recommended at the end of *PA* I 1, which says that when inquiring into a part one should seek both the end for the sake of which the part exists and the necessary causes that the formal nature uses to bring about that end. Leunissen takes the whole of *PA* III 2 to be a discussion of secondary teleological causation. However, if I am right that *PA* 663b20–4 expresses the methodological norm at the end of *PA* I 1 – which tells us that *first and foremost* inquiry should be carried out in this way if we want to grasp the causes – then it is a mistake to restrict these double-barrelled explanations

[46] For my discussion of this example see Chapter 8. It is not clear how Case 4 fits into Leunissen's categories (she does not appear to discuss it). On the one hand, these particular membranes are not 'subsidiary and luxury' parts like the omentum but vital for the organism's survival. This should make them products of her primary teleology according to the (B) part of the matrix above. Moreover, the example appears to illustrate the teleological principle behind *GA* 743a36–b18, which Leunissen agrees underwrites her category of primary teleology. On the other hand, Leunissen (2010, 201) associates this kind of 'double-barrelled' explanation with her category of secondary teleology. In that case, the example should be read as describing how nature *co-opts* the membranes that arise through material necessity alone in order to satisfy an emerging biological need that is not specified in advance by the pre-existing potential for form. But then we lose the connection between secondary teleology and the production of subsidiary/luxury parts: extraembryonic membranes do not just enhance the organism's flourishing but are conditionally necessary for its survival. This difficulty speaks to my main concern about trying to box Aristotle into a commitment to two discrete and mutually exclusive categories of teleology that are supposed to apply to two different kinds of part.

to a subsidiary form of secondary teleology (Leunissen 2010, 201–2); they are the core of Aristotle's natural teleology.

Finally, we can see this causal pattern reflected in our other key passage at *GA* 743a36–b18, which we have seen applies to the formation of each of the uniform parts: 'We must state then, as we have said, that *all of these things* come into being, on the one hand, by necessity, and on the other hand, not by necessity but for the sake of something.' It is hard not to read this as an expression of the same norm of inquiry that we have been considering. And as this passage makes clear, success here depends on being able to identify not only the function of each part but how the formal nature of the developing organism 'uses' material necessity as a co-cause in realising that end.

It is worth noting how the *PA* I 1 approach to causal inquiry where we attempt to show how some feature of a living thing comes to be both from necessity and for the sake of an end has close ties with the methodological norms of the *Timaeus*.[47] There, too, Plato recommends that inquiry into nature should endeavour to uncover both necessary and final causes:

> So anyone who is a lover of understanding and knowledge must necessarily pursue as primary causes those that belong to intelligent nature, and as secondary all those belonging to things that are moved by others and set still others in motion of necessity. We too, surely, must do the same: we must describe both types of cause. (46d8–e3, Zeyl translation with modifications)
>
> That is why we must distinguish two forms of cause, the divine and the necessary. On the one hand, we must search (*zêtein*) in all things for the divine cause if we are to gain a life of happiness to the extent that our nature allows, and on the other hand, for the necessary cause for the sake of the divine. (68e6–69a2, translated after Zeyl)
>
> We have now sorted out the different kinds of cause, which lie ready for us like lumber for carpenters. From them we are to weave together the remainder of our account. (69a5–7, Zeyl translation)

If, as I argued in Chapter 8, Aristotle inherited the architectonic model of generation from the *Timaeus*, then it should not surprise us to find him recommending a similar method in *PA* for inquiring into living things and their parts.

[47] This has also been noted by Strange (1999 n. 10).

CHAPTER 9

The Cosmological Significance of Substantial Generation

In Chapters 5 and 6 we saw that Aristotle treats male and female as 'principles' (*archai*) of generation both in the sense of starting points (animals come to be from the union of male and female) and in the sense of causes (the male as the efficient cause, the female as supplying the material cause).* However, Aristotle does not treat male and female as causally basic principles in the sense of being causes of other things while nothing more fundamental is the cause of them (cf. *GA* 788a14–16). At the beginning of *GA* II 1 Aristotle tells us that the existence of males and females themselves can be traced to a higher principle (731b24–5: *anôthen*) namely, the divine. And so, they are not the *most basic* principles of animal generation since something more fundamental is the cause of them. In this chapter I will explore the argument at the beginning of *GA* II 1 as a way of examining the place of substantial generation in Aristotle's broader cosmology. Although most of what I have to say focuses on the generation of biological substances, my interpretation will apply equally to cycles of elemental transformation.

9.1 Two Perspectives on Generation

David Sedley is perhaps the most well-known defender of the cosmological interpretation of Aristotle's teleology. According to Sedley, Aristotle thinks of the universe as an organised whole whose parts (elements, animals, and plants) are all coordinated in such a way that their mutual interactions contribute to the good of the cosmos and, ultimately, the good of man. On

* This chapter arose from a paper given at the Princeton Classical Philosophy Colloquium (7 December 2013), which then became Henry (2015b). The current chapter is a revised version of that published paper. For some interesting variations on the themes of this chapter and a more detailed discussion of the background issues see Quarantotto (2015). As will become clear, I am generally sympathetic to Quarantotto's interpretation though I disagree with some of the details of her arguments.

this reading, Aristotle's natural teleology cannot be fully grasped in isolation of his cosmological theory.[1] In opposition to this, many modern scholars defend what I will call the "organism-centred" view of Aristotle's natural teleology. This reading treats the individual organism as the final end for the sake of which all its features exist, so that all final causation in biology can be understood exclusively from the perspective of the good of the individual organism itself. On this reading, each individual has the features it does because, and only because, those features are good for that individual.[2]

I count myself among those who accept some version of the organism-centred view. For the most part Aristotle treats the survival and well-being of each individual organism as the final end for the sake of which its features ultimately exist. And while those features might also benefit things outside the individual organism, those benefits are not part of the causal story that explains why they are present in the first place. At the same time, I think there is value in considering what Quarantotto (2015) calls 'the view from above' insofar as it forces us to take seriously the connections between Aristotle's natural teleology and his broader cosmology.[3] In my view the collective reaction to Sedley's interpretation has had the effect of swinging the pendulum too far in the opposite direction. As a consequence, scholars have tended to ignore (or at least downplay) the cosmological features of Aristotle's biology. To be sure Sedley overestimates his case. For, as far as I can tell, there is only one exception to the organism-centred view (discussed below). And so it is hardly enough to vindicate the Cosmic Teleology reading. For the explanation in question does not form part of any systematic attempt by Aristotle to adopt a more global teleological perspective of the sort defended in Plato (e.g. *Laws* X, 903b5–d1, *Phaedo* 98b1–4). However, I think this passage is enough to show that Aristotle

[1] Sedley (1991, 180; 2010, 24).

[2] For a classic statement of this view see Balme (1972, 96–7). It is defended by Bodnar (1997), Lennox (2001a, e.g. 182–4; 2001b, 341), Judson (2005, 348, 359–60), Gotthelf (2012, e.g. 8–9 n. 13), and Leunissen (2010, e.g. 41: 'cosmic teleology is ultimately grounded in natural teleology' and 'natural teleology in its primary form is limited to individual natures acting as efficient causes for the sake of something'). *Physics* II 7, 198b5–9 is among the main passages cited in defence of the view that, for Aristotle, each organism's nature aims at its own good. Lennox (2001b, 341) calls this Aristotle's 'basic teleological axiom'. Not everyone who rejects cosmic teleology accepts the organism-centred view. Johnson (2005, 278), for example, argues that each particular substance is the final end for the sake of which all its features exist, so that all final causation can be understood exclusively from the perspective of the particular substance itself. But he denies that Aristotle thinks organisms are the only natural substances subject to teleological explanation. The elements (on his reading) also count as substances.

[3] Scharle (2015) calls this the 'cosmic perspective'. While I am sympathetic to Scharle's view, I do not see this cosmic perspective as being nearly as thoroughgoing in Aristotle as she does.

held a weaker version of the organism-centred view than has traditionally been ascribed to him by Sedley's opponents.

The exception I have in mind comes at the beginning of *Generation of Animals* Book II, where Aristotle is attempting to give an account of the final causes of sexual differentiation. I will argue that Aristotle's argument here is a specific application of a more general argument he gives in *GC* II 10, which seeks to explain why natural generation occurs continuously and without fail by placing it in the context of his broader cosmology. What is at stake here is exactly how to understand the teleological significance of natural generation. If I am right, then what Aristotle is arguing in *GA* II 1 is that continuous animal generation exists not (or not only) because of the contribution it makes to the individual's own good but because of its contribution to the good condition of the universe as a whole. But before turning to that let me begin by drawing attention to another passage to give a sense of what I have in mind by "the cosmological features" of Aristotle's biology.

In *GC* II 10 Aristotle describes how the sun's annual motion around the earth causes generation and destruction in the sublunary world as it approaches and retreats. Part of that story includes an account of how this motion influences the life-cycles of living things:

> This explains why it is not the primary motion that causes generation and destruction, but the motion along the inclined circle [sc. the sun's annual motion in the ecliptic] . . . the approaching and retreating of the moving body are caused by the inclination. For the consequence of the inclination is that the body becomes alternately remote and near . . . Therefore, if it generates by approaching and by its proximity, it – the very same body – destroys by retreating and becoming remote . . . Hence, too, the times (i.e. life-cycles) of the several kinds of things have a number by which they are distinguished. For there is an order for all things, and every time (i.e. every life-cycle) is measured by a period. Not all of them, however, are measured by the same period, but some by a smaller and others by a greater one. For some of them the period that is their measure is a year, while for others it is longer and others shorter than this. (*GC* II 10, 336a33–b16, Joachim translation with modification; cf. *GC* II 11, 338b3–5)

In *GA* IV 10 Aristotle explains how this works in the case of animal generation (777b16–778a9). There he tells us how the lives of organisms 'strive to be measured according to natural cycles' (i.e. days, months, and years), which are determined by the motions of the sun and moon:

> For the month is a common cycle between these two [sc. the sun and moon], and the moon is a principle owing to its communal relationship with the sun

> and its participation in its light; for it becomes just like a lesser sun. For this reason the moon contributes to all processes of development and completion. For the heatings and coolings, in a certain proportion, produce generation followed by destruction. And the limits of these (both their starting points and terminating points) are contained by the motions of the stars. For just as we observe a sea and everything that has a fluid nature being stabilised and changed according to the motion and stillness of the winds, and the air and the winds according to the cycles of the sun and moon, so too both the things that grow from these and other sorts of things necessarily follow upon them. For it is rational for the cycles of the less important things to follow upon those of the more important things. For even wind has a kind of life, a generation and destruction. And there are probably certain other sources of the revolutions of these stars. Therefore, nature seeks to count out the processes of development and completion by the numbers of these. But it is not precise owing to the indeterminacy of the matter and because of the many principles that impede the natural processes of generation and destruction and frequently cause things that happen contrary to nature.

The explanation here has two parts. The *efficient causal* explanation attempts to identify the mechanism by which the stages of an animal's life-cycle are ultimately controlled by the motions of the sun and moon. Their motions cause changes of the seasons, bringing with them changes in temperature and light, which in turn trigger events within the organism associated with the phases of its life (e.g. as the days get shorter there is a decrease in light and temperature that triggers the hibernation instinct in certain animals). But Aristotle also thinks the phenomenon in question can be given a *teleological* explanation. Aristotle is not suggesting that the cosmic cycles occur for the sake of regulating the phases of an animal's life: they do not occur *in order that* animals should have something by which to measure their life-cycles. Instead he is saying that the formal natures of animals *use* the motions of the stars and the changes that follow upon them for that purpose. And being "used for the sake of something" is a kind of teleological relation.[4]

I draw attention to this passage in order to illustrate how Aristotle thinks that the full significance of certain aspects of animal generation cannot be grasped without reference to cosmology. For the explanation on offer depends, at least in part, on understanding Aristotle's broader theory about the causal dependence of changes in the sublunary world on the eternal motions of the heavens that he articulates in *GC* II 10. Of course, in

[4] See Leunissen (2010, e.g. 81–99).

this particular example the appeal to teleology does not involve relating the feature in question to some broader cosmic good, and so this observation is entirely compatible with the organism-centred interpretation. Thus, while the *GA* IV 10 passage illustrates *the cosmological significance* of animal generation, it does so without depending on *a global teleological perspective* of the sort we find in Plato. But the same cannot be said for the *GA* II 1 passage that forms the focus of this chapter. In that passage Aristotle (I claim) adopts the global teleological perspective in order to explain the feature in question.

It is worth stressing at the outset that the arguments in this chapter should in no way be taken as a defence of Cosmic Teleology. I think that reading fails as an overall interpretation of Aristotle.[5] Instead my aim is to advocate for a slightly weaker version of the organism-centred view than opponents of Sedley have traditionally defended. While it is true that, for the most part, Aristotle treats the organism's own good as the final end for the sake of which its adaptations ultimately exist, I want to stop short of claiming that *all* final causation in Aristotle's biology can be understood *exclusively* from the organism's perspective. Instead I shall attempt to show that there is at least one feature of the natural world – the continuous process of substantial generation itself – that cannot be understood from the perspective of the individual substance alone but must be explained (at least in part) by the contribution it makes to the universe as a whole.

9.2 Why Are There Sexes?

One of the major questions of the *Generation of Animals* is why sexes exist. This question has two parts (1) Why are there sexes at all? That is, why do living things possess male and female principles? (2) Why are these principles separated in animal species? That is, why are animals (unlike plants) divided into separate sexes? Aristotle thinks that plants contain male and female principles, but these are present together in the same individual. So there must be some explanation as to why they are divided in animal species. *GA* II 1 opens by introducing two modes of explanation for (2): one that refers to the embryological mechanisms responsible for actually differentiating individuals into males and females (efficient causation), and one that refers to the advantage that comes from having separate sexes (final causation).

[5] I defend this in detail in Henry (2015b).

> We have said before that female and male are principles of generation and what is the capacity that defines the substance of these (*hê dunamis kai ho logos tês ousias autôn*). In what follows we shall attempt to give an account of why one individual comes to be female and another male both as coming to be from necessity (i.e. from what primary mover and what sort of matter) and because of the better and the cause for the sake of something, which is the higher principle (*anôthen echei tên archên*). (731b18–24)

The opening argument of *GA* II 1 provides the teleological explanation for the separation of the sexes while the efficient causal explanation is postponed until Book IV 1–2.[6] In this chapter I will focus exclusively on the teleological argument. The argument contains two parts. The first part answers the question of why sexes exist at all, while the second part (what I shall call the Separation Argument) contains Aristotle's answer to the question of why these principles are divided into male and female individuals in the case of animals. My main focus in the chapter is on the first part, but I begin with the second.

9.2.1 The Separation Argument

Aristotle's argument for the separation of the sexes is a continuation of the discussion at the end of Book I. In *GA* I 23 Aristotle tells us that in all animals that are capable of progressive motion the male and female principles are separated into different individuals, while in plants these two capacities are present in the same individual (730b32–731a4). In one of the most striking images of the entire treatise Aristotle likens animals to a plant that has been 'torn asunder, as if one were to pull apart and separate it into the male and female contained in it when it was bearing seed' (731a21–4), though they reunite again like a plant for the sake of generation (b5–8). But why, if the male and female principles have to be united in order to accomplish the task of generation, would nature tear them asunder in the first place? The answer to this is supplied by the Separation Argument, which purports to show why in animal species 'one individual comes to be female and another male'.

That argument depends on the principle, articulated back in *GA* I 4, that nature does everything *either* because it is necessary for an end *or* because it is better (717a15–17).[7] While the possession of male and female principles is necessary for generation, their separation cannot be; for plants can generate

[6] For a discussion of the mechanism of sexual differentiation see Henry (2007a, 2006a, 2006b) and Connell (2016, chapter 9).

[7] For a discussion of this principle see Leunissen (2010, 125ff.).

without having those powers in separate bodies. Therefore sexual differentiation must somehow make generation better ('because of the better and the cause for the sake of something'). Aristotle's challenge, then, is to show why having these principles in separate bodies is better for animals. The answer to this will supply the final cause of the phenomenon. The argument runs as follows:

> Since the principles of these [sc. the *genos* of humans, animals, and plants] are male and female, male and female will be present for the sake of generation in each of the things that possess them. But the primary moving cause is better and more divine in its nature than the matter, insofar as the definition and the form belong to it, and it is better that the superior cause be kept separate from the inferior one. It is on account of this that (in those species where this is possible) the male is separated from the female. For the source of change, to which the male principle belongs, is better and more divine in those things that come into being, while the female corresponds to matter. However, the male comes together and combines with the female in order to perform the function of reproduction, for this is something common to both. (732a1–12)

The argument in this passage has the following structure:

(1) Wherever possible, it is better to have the superior principle separate[8] from the inferior principle.
(2) The male principle corresponds to the primary moving cause, while the female principle corresponds to the material cause.
(3) The primary moving cause is superior to the material cause.
(4) Therefore, the male principle is superior to the female principle.
(5) Therefore, whenever possible, it is better to keep the male principle separate from the female principle.

Aristotle does not say why it is better to keep the superior male principle separate from the inferior female principle. Some commentators supply the missing justification from *GA* I 23, where Aristotle mentions the fact that animals differ from plants by possessing sense perception:

> In all this nature acts like a sensible craftsman. For the essence of plants includes no other function or business aside from the production of seed. Since, then, this is brought about by the union of male and female, nature has mixed these and set them together in plants, so that the sexes are not divided in them. Plants, however, have been investigated elsewhere. But the

[8] I take it that by "separate" Aristotle means separate in magnitude or place (i.e. existing in different bodies) and not just separate in account or definition. For the male and female principles are separate in account in plants (see *GA* I 20, 729a24–34).

> function of the animal is not only to generate (which is common to all living things), but all of them also partake in a kind of knowledge, some more, some less, and some very little indeed. For they have sense-perception, and this is a kind of knowledge . . . Now it is by sense-perception that an animal differs from those organisms which have only life. But since, if it is a living animal, it must also live, when it is necessary for it to accomplish the function of that which has life, it unites and copulates becoming unified like a plant again, as we said before. (GA I 23, 731a25–b7)

David Balme reads this passage as making the claim that sexual differentiation exists in animals, not for the sake of reproduction (a function they share with plants), but for the sake of cognition. On this reading the Separation Argument is meant to show that males can engage in cognitive activity better if they do not contain any trace of the female principle: 'Aristotle completes this argument in the first section of Book II by saying that the male is superior, and that it is better for the superior to be separated from the inferior – presumably because the male can function better as a cognitive animal when not combined with the more material female nature (732a3)' (Balme 1987, 153; see also Connell 2016, section 8.6).

This interpretation (call it the cognitive efficiency reading) is not implausible. In *PA* II 4 Aristotle connects differences in blood to differences in sensory capacities. And he thinks the male principle differs from the female principle in being hotter and stronger (i.e. more effective at concoction). It is not unreasonable to suggest that this difference in turn translates into a difference in their blood in ways that affect perception. In that case Aristotle could be saying that it is better *for perception* if the hotter male principle is kept separated from the colder female principle, since blood that has been concocted by a male principle alone will be hotter and therefore more conductive to perception than blood that has been tempered by the cooling and solidifying effects of the female principle. Even then I am not convinced Balme's interpretation is right. Aristotle does not say in *GA* 732a1–12 that it is better *for perception* if the sexes are kept separate let alone that they are separated so *males* can have a better perceptual experience. Nor does he say this in the *GA* I 23 passage. A careful reading of *GA* I 23 shows that the emphasis throughout is on the union of the sexes and its relation to generation. The point about animals there is that, even in these cases where the sexes exist in distinct individuals, we see them coming together for the sake of reproduction: 'For when they need to reproduce they become un-separated (*achôriston*), just as in plants, and the nature of each wishes to become one. This is clear according to perception where a kind of unified animal comes to be from both when the individuals are united and coupling' (*GA* I 23, 731a9–14).

If we read 731a25–b7 carefully, we can see that the emphasis is still on the union (or un-separatedness) of the sexes. Aristotle raises the point about sense perception there only to say that animals are different from plants in that way. But then he immediately points out that, despite this difference, because an animal is also a biological organism it must also participate in the most basic functions of life, which is to reproduce another like itself (731b5–8; cf. *DA* II 4, 415a22–5). And so 'when it is necessary for it to accomplish the function of that which has life, it unites and copulates becoming unified like a plant again'.[9] If this is right, then Aristotle does not invoke sense perception to explain why the sexes are separated in animals. For that is not the focus of *GA* I 23.

Having said that, I do think Aristotle's remarks about sense perception at the end of *GA* I 23 are relevant to the Separation Argument.[10] Premise (1) of that argument says it is better to separate the superior principle from the inferior principle *wherever possible*. Observation shows us that male and female principles must be united in order to execute the function of reproduction. Since plants lack the sensory part of the soul, which houses the capacities necessary to navigate their environment and locate a mate, if they were separated into males and females they would not be able to come *back* together again and unite for generation. (Obviously Aristotle was unaware of the mechanism of cross-pollination by insects!) And so it is not possible to separate the male and female principles in the case of plants and still accomplish the basic functions of reproduction (cf. *GA* I 1, 715a26–30).[11] By contrast sexual differentiation is possible in the case of animals since they possess the powers of sensation and locomotion. (Notice that *GA* I 23 opens by saying that the sexes are separate in all those animals that have the power of progressive motion.) On this reading sensation is not the final cause of sexual differentiation, as it is on the cognitive efficiency interpretation, but simply the feature that makes sexual differentiation possible.

Armed with the *GA* I 23 point in hand we can fill out the Separation Argument a bit more:

[9] Even the image of animals as plants with their sexual principles torn apart strikes me as focused on the fact about the sexes being united for the sake of generation rather than their separation.

[10] I owe this reading to an unpublished paper written by Sean Coughlin, though it was originally suggested by Balme (see his comments on *GA* 715a26).

[11] Alternatively, Aristotle might be thinking that since plants lack sensory and locomotive powers sexual differentiation would be in vain, and nature does nothing in vain.

(1) Wherever possible it is better to have the superior principle separate from the inferior principle.
(2) The male principle corresponds to the primary moving cause, while the female principle corresponds to the material cause.
(3) The primary moving cause is superior to the material cause.
(4) Therefore, the male principle is superior to the female principle.
(5) Therefore, whenever possible, it is better to keep the male principle separate from the female principle.
(6) It is possible to separate these principles in animals that are capable of progressive motion.
(7) Therefore, the male principle ought to be separate from the female principle in those animals.

Aristotle appears to have empirical grounds for accepting Premise (2). On the basis of observations concerning the differences between semen and menstrual blood, Aristotle inferred that males are hotter and stronger than females, where hotter and stronger are again measures of a thing's ability to do work (in this case the power of its heat to effect concoction). This provided him with good reasons for identifying the male principle with the primary efficient cause of change just as rennet is the agent that causes milk to coagulate in virtue of the natural heat it contains (739b20–6).[12] Premise (3) follows from the causal principle that what is potentially *F* (the matter) can only be made into an actual *F* by what is already an *F* in actuality. It follows from this that in order to generate a thing the primary efficient cause must already possess the form of an *F*, not potentially, but in a state of completion. And since matter, which exists potentially, derives its substantial being from the form, which is its end and actuality, the form is ontologically and teleologically prior to the matter (*Metaph.* IX 8, 1050a15–23, b2–3) and in that sense is superior to it. And if it is better, whenever possible, to keep the superior principle separate from the inferior principle, it follows that in animal species the male principle ought to be separated (not only in account but also in place) from the female principle. And nature does everything either because it is necessary for some end or because it is better.

But we might still wonder: better *for what* (or *for whom*)? Aristotle does not give a satisfactory answer to that question. Elsewhere I suggested that the Separation Argument presupposes an ideology in which males are

[12] For a reconstruction of the empirical case see Henry (2007a), Mayhew (2004, chapter 3), Connell (2016, 270–84).

viewed as better and more divine than females (an ideology clearly endorsed at *GA* 716a14–18; cf. *Ph.* 192a17–25).[13] There I argued that what Aristotle really intends the argument to demonstrate is not just why animals are divided into sexes but why the *superior male* principle is kept apart from the *inferior female* principle. In doing so nature is not merely attempting to separate the two sexes; the aim is to avoid the contamination of the one by the other. This suggests that "better" is being used in the unqualified sense. In that case Aristotle is saying that keeping the superior principle separate from the inferior principle is better *simpliciter* – not better *for* anything. Premise (1) is a basic normative assumption of Aristotle's metaphysics just as the claim that up is better than down, right than left, living than non-living, soul than body, and being than non-being.[14] The virtue of this reading is that it makes the argument work as presented in the text without the need to fill in any unstated premises.

9.2.2 The Cosmological Argument

As we have seen, the Separation Argument presupposes the existence of male and female principles and asks why, in the case of animal species, they are present in separate bodies. But why should sexes exist at all? What is the cause on account of which living things contain male and female principles in the first place? Aristotle's explanation for this refers the existence of the sexes to a 'higher principle' (*anôthen echein tên archên*), namely, the divine:

> [A] Of the things that are, some are eternal and divine, while others are capable of being and not-being, and the beautiful and divine is always a cause in virtue of itself of what is better in things that admit it. And what is not eternal is capable of being and admits of the better and worse. And soul is better than body, and what is ensouled is better than what is not owing to the possession of soul, and being is better than not being, and living is better than non-living. These are the causes owing to which the generation of animals exists. [B] For, since it is not possible for the nature of this kind of thing to be eternal, what comes into being is eternal in the only way possible for it. While it is not possible to be eternal *in number* (for the substance of existing things is in the particular; if it was of this sort, then it would be eternal), it is possible to be so *in form*. For this reason there is always a continuous generation[15] of humans, animals, and plants. [C] Since

[13] See Henry (2007a, 16–18).
[14] *PA* 648a11–12 (notice the reference to male and female here), *PA* 672b23, *GA* 731b28–30.
[15] I follow Lennox in taking *genos* here to have the sense given at *Metaph.* V 28, 1024a29–30: 'the continuous generation of things having the same form' (what I shall call an ancestor–descendant lineage).

> the principle of these is the male and female, it is for the sake of generation that the sexes are present in those that possess them. (*GA* II 1, 731b24–732a3)

Text [A] starts by dividing existing things into those that are everlasting and divine, on the one hand, and those sublunary things that are capable of both being and non-being, on the other. We have already encountered this distinction in connection with *GC* II 9, where Aristotle also distinguishes between the 'eternal and primary things' and those sublunary beings that are such as to come to be and pass away. That discussion goes into more detail about the difference between the two kinds. The relevant text comes at *GC* 335a34–b5:

> For some things exist from necessity (e.g. eternal things) while others necessarily do not exist. Of these, because it is not possible for them to be otherwise than in their necessary state, the first are incapable of not existing while the latter are incapable of existing. But other things are capable of both existing and not existing, which are precisely those that are generable and destructible. For this type of thing exists at one time and does not exist at another. Therefore it is necessary for generation and destruction to occur among those things that are capable of existing and not existing.

Generable (sublunary) substances are characterised here by a sort of temporary being insofar as each one 'at one time *is* and at another time *is not*' (*pote men gar esti touto, pote d' ouk estin*, 335b3–4).[16] The cause of this is that sublunary things are composed of matter which is the seat of their capacity to-be-and-not-be (335a33). It is in this sense that matter was said to be a principle and cause (a32) of generation and destruction. Aristotle then contrasts these with the class of necessary things: both those that exist from necessity (e.g. the eternal and divine things) and those that necessarily do not exist (*ta d' ex anangkês ouk estin*) (e.g. a square circle). Since it is not possible for any of these to be otherwise than in their necessary state (*to mê endechesthai para to anangkion allôs echein*) the first kind cannot fail to exist while the second are incapable of existing. As such neither of these are generable or destructible. It is only those things that are capable of being and not being (owing to the presence of matter) that come to be and pass away. In light of this I propose we take Aristotle's point in [A] as follows.

All existing things partake in being but do so to different degrees. Divine substances enjoy being eternally, since they exist from necessity and thus

[16] For my interpretation of this passage see Chapter 4. I take Aristotle to use *estin* here in its complete and existential sense. However, as Shields (2007, 141) notes, substantial being for Aristotle is not only the cause of a thing's existing (being in the existential sense) but also the cause of its being the kind of substance it is (the sortally determinate sense). This is the point we took away from *DA* 415b12–14.

are incapable of not existing. By contrast sublunary substances enjoy being but do so only intermittently (this type of thing exists at one time and does not exist at another). For they contain matter which is a capacity in virtue of which things are capable of being and not-being. When Aristotle then says that 'what is not eternal is capable of being and admits of the better and worse' we can take him to mean sublunary substances are nevertheless capable of enjoying being eternally to varying degrees of success.

Text [B] then expands on this point. While no sublunary organism is capable of existing indefinitely (since each one is composed of matter and so eventually decays), they can become eternal in form by engaging in reproduction. The upshot of this, Aristotle says, is that while there cannot be eternal *organisms* in the sublunary realm there can be eternal *lineages* ('For this reason there is always a continuous generation of humans, animals, and plants.'). This, in turn, provides the final cause for the existence of males and females: [C] male and female are principles of generation and as such are (conditionally) necessary for maintaining those lineages, which are in turn the means by which sublunary substances achieve eternal being.[17]

Traditionally this passage has been read simply as a version of the argument at *DA* II 4, 415a26–b7. There Aristotle attempts to explain why organisms ('those that are complete and not maimed or spontaneously generated') do everything they do for the sake of reproduction ('in order to produce another like itself') by tracing it to a natural desire 'to partake in the everlasting and divine'. Since no living thing can endure as numerically the same individual (all material composites inevitably decay), the next best thing is for it to produce something that is like it in form.[18] Most scholars take Aristotle to be making the same point in *GA* 731b24–732a3. On this reading, the perpetuation of the species is a mere consequence of the fact

[17] See Quarantotto (2015, 164–5). The one aspect of Quarantotto's account that I disagree with is what she calls Aristotle's 'gradualist perspective' on being. On her reading divine things enjoy eternal being while sublunary things strive for this being but only achieve an approximation of it insofar as they partake in an everlasting process of change. My way of reading the relevant texts is subtly different. On my reading the distinction is not between those things that partake of *eternal* being and those that only approximate it but between those things that partake of being *eternally* and those that do so *only intermittently*. *GC* II 9, 335a34–b5 seems to be making that point. When Aristotle says there that perishable things are capable of both being and non-being he means that this type of substance exists at one time and does not exist at another (335b3–4). What he seems to be saying here is that each particular perishable substance enjoys (and not simply approximates) full being *while it exists* but that it does so *only intermittently*, whereas divine things enjoy being eternally (since they cannot *not* exist). Thus, Aristotle is not saying that *the being* of the various classes of existing things differs by the more and the less. Rather all substances partake of (full) being but do so to varying extents: some always partake of being while others do so only intermittently.

[18] Compare *Symposium* 206a–209e.

that, when every individual organism strives after its own good (personal immortality), what results is a continuous generation of things of the same form. But that is not the end for the sake of which reproduction exists.[19] I will call this the organism-centred reading of the *GA* argument, since it treats the good of the individual as the end for the sake of which generation exists while species (eternal lineages) are a mere byproduct of this.

While the argument at *GA* 731b24–732a3 obviously has some parallels with *DA* 415a26–b7, scholars have been too quick to assume that they are the same argument. But there is another possible reading that I think is more interesting. I want to argue that *GA* 731b24–732a3 should instead be read as a specific application of a more general argument that Aristotle makes in *GC* II 10, which seeks to explain why coming-to-be and passing-away occur perpetually and without fail. What is at stake between these two readings is exactly how to understand the teleological significance of biological generation itself. If I am right, then what Aristotle is arguing in *GA* II 1 is that *continuous* generation exists not (or not only) because of the contribution it makes to the individual's own good but because of its contribution to the good state of the universe as a whole.[20] In this way *GA* II 1 not only highlights the cosmological significance of animal generation by adopting 'the view from above' (as Quarantotto puts it); it attempts to do so by explaining the phenomena from a global teleological perspective. To see this, we need to examine *GC* II 10 in more detail.[21]

19 Kullman (1985, 172) writes of the *GA* II 1 passage: 'The idea is, once again, that the circle of procreation is the way in which animals, inasmuch as they are sexual, partake of the everlasting and the divine, which in this respect functions as the final cause.' See also Balme (1991, 96; 1987, 279–80), Lennox (1985, e.g. 67, 73, 92 n. 9), Leunissen (2010, 41), Gotthelf (2012, 9 n. 13, 58, 71 n. 13), Shields (2016, 210). Quarantotto (2015, e.g. 164) also lumps *GA* II 1 together with *DA* II 4 in opposition to *GC* II 10. Johansen (2015) offers a similar assessment of the relation between the *GA* and *DA* arguments, but he takes the latter to hold that reproduction exists for the sake of preserving the species. I don't think the *DA* argument supports that reading (though *GA* 767b8–10 comes close).

20 Segev (2015) criticised an earlier version of this argument (see Henry 2015b) on the grounds that it suggests the universe is somehow the ultimate beneficiary of sexual reproduction. The way I put the point there – that continuous animal generation contributes 'to the good of the universe' – certainly invited that criticism (though I'm not sure Segev's alternative is any less objectionable). The present formulation of the argument is more careful to distinguish what is good *simpliciter* from what is good *for* something. On my reading, Aristotle thinks it is good *that* the universe contains more being than non-being (good *simpliciter*). In *GC* II 10 he claims that perpetual generation exists for the sake of its contribution to that good state of affairs. But this does not commit Aristotle to the view that there must be some beneficiary of that good, i.e. that it is good *for* the universe that it contains more being.

21 For a similar reading see Cooper (1982). Cooper recognises that *GA* II 1 treats the continued existence of living things itself as an important good and not just the existence of the individual organism. But he does not trace the argument back to *GC* II 10 in the same way. By contrast, while Gotthelf (2012, 58) sees the connection between *GA* II 1 and *GC* II 10, he continues to treat the former as if it were

9.3 The Continuity of Generation: *GC* II 10

GC II 10 asks: what is the cause responsible for the fact that generation and corruption go on continuously and never fail? He has already provided the material cause of the phenomenon back in *GC* I 3 (see 319a17–22). *GC* II 10 offers the efficient and final causes. The efficient cause explanation comes at 336a15–336a7. Back in the *Physics* Aristotle had established that locomotion is the primary form of change and that this is what causes generation and destruction. He fills out the picture in *GC* II 10 by identifying the motion in question with the sun's annual movement in the ecliptic as it approaches and retreats from the earth's surface (cf. *Meteor.* 346b20–347a6). This, Aristotle says, can be supported by appealing to the observable facts: 'for we see (*horômen*) that generation occurs as the sun approaches and things decay as it retreats' (336b16–24). Since the sun's motion is eternal, it follows that generation and destruction must themselves occur continuously and without fail.

GC 336b25–337a7 then provides the teleological explanation for the continuity of generation and destruction in the sublunary realm by connecting it to "the better". The main text of the argument runs as follows:

> As we have said, coming to be and passing away will be everlasting and continuous and will never fail owing to the reason stated. This conclusion also follows rationally (*eulogôs*). For, since we say that nature always strives after what is better and that being is better than not being (we said elsewhere that "being" is spoken of in many ways) but that it is not possible for this to belong to all things on account of their being far removed from the first principle, the god filled up (*suneplêrôse*) the universe with being in the way that was left by making generation perpetual. For that way being would be most connected (*malista suneiroito*) because coming to be eternally is the closest thing generation has to being. (*GC* II 10, 336b25–35)

Two preliminary remarks are in order here. First, we can set to one side the fact that this passage invokes the rather *un*-Aristotelian claim that the universe was created by a providential god who deliberated about what is best for the world. One possibility is that Aristotle is simply invoking the popular idea of a creator god as a heuristic device. Reasoning *as if* the universe was designed by a providential god aiming at some optimal state of affairs is a useful heuristic device because it helps us identify the final cause of continuous generation. Aristotle, like Plato, takes it as a basic fact, not explicable in terms of anything more fundamental, that the universe is an

simply a version of the *DA* II 4 argument. On his reading, the lesson of *GA* II 1 is that reproduction is aimed at *self*-preservation, while the existence of eternal lineages is merely a consequence of that.

ordered system and that this order is good. It follows that everything that contributes to this order is itself good. But that does not commit Aristotle to the further claim that there must have been some agent (e.g. the Prime Mover) who organised the parts of the universe in order to make it so.[22]

Second, when Aristotle talks about coming-to-be in this passage he undoubtedly means coming-to-be in the unqualified sense and so "being" should be understood in its complete and existential (rather than incomplete and predicative) form.[23] Now I think Quarantotto (2015, 164) is right that the emphasis here is on *everlasting* being. For example, when Aristotle says that 'it is not possible for this [sc. being] to belong to all things' he is not referring to the fact that some things are incapable of being at all (cf. 335a34–5). Rather the distinction is again between those things that enjoy being *eternally* (divine substances) and those that enjoy being *only intermittently* (sublunary substances). Members of the former kind cannot fail to exist since they exist from necessity, while each of the latter exists at one time and does not exist at another. This is what is important about continuous natural generation. It connects things that exist only temporarily into a process that 'imitates' circular motion (the only truly eternal motion) and thus allows sublunary being to continue perpetually and without fail.[24]

The conclusion of the argument is fairly clear. Continuous natural generation exists because it is the best possible way to maximise the level of being in the universe as a whole given the constraints on sublunary being. Aristotle deduces this conclusion from two universal principles, both of which are taken as basic:

[22] I owe this suggestion to Sean Coughlin. On the use of teleological principles as heuristic devices see Leunissen (2010, 119–35).

[23] Recall that when used in the qualified sense "*S* comes to be" remains incomplete without specifying what the subject comes to be (*S* comes to be *F*), e.g. we say that a man comes to be *healthy* or comes to be *bigger*. When used in the unqualified sense "*S* comes to be" is complete all on its own (*S* comes into being, full stop), e.g. we say that a human has come to be that did not exist before.

[24] While I disagree with Quarantotto's gradualist perspective on being (see note 17 above), she is right to point out that what is important about the Prime Mover is not its unchangingness but its everlastingness and that everything else in the cosmos imitates its everlastingness (not its unchangingness) through change. In this sense Aristotle's ontology is (as she puts it) radically dynamic in that the natural world accomplishes perpetual being by means of perpetual change (2015, 185). Cooper (1982, 204) also emphasises the eternal nature of the sublunary world as key to Aristotle's natural teleology and takes it to be a basic assumption that grounds that theory (one that is not explicable in terms of any more fundamental facts but is instead justified on observational grounds). However, this feature of the sublunary world is explained in *GC* II 10 by appeal to final causes, specifically to principles of optimisation (that a world filled up with being is better than one with existential gaps). This in turn explains the fact that sublunary generation and destruction are eternal and continuous. See below.

(i) Nature always strives after what is better.
(ii) Being is better than non-being.

The first of these is explicitly identified as a first principle of natural science in *IA* 2, which states that 'nature does nothing in vain but always what is best for the substance from among the possibilities concerning each kind of animal' (704b14–15).[25] This assumption is supposed to license a certain pattern of explanation that appeals to the concept of what is best (what I call "optimality" reasoning). According to this pattern of reasoning, we explain why something is the way it is by showing how being in that state is *the best possible* way for it to be. As Aristotle puts it, 'if it is better this way, then it is that way [sc. since nature always strives after what is better] and being in that state is in accordance with nature' (*IA* 704b17–18).[26]

The teleological argument in *GC* II 10 exhibits this pattern of reasoning. Since being is better than non-being, the ideal universe would be one that contained only eternal substances that never passed out of existence. That way the universe would be completely filled up with being with no traces of non-being. But since the actual universe contains a sublunary realm of material composites, and since all material composites must come to be and pass away (*GC* II 9, 335a32–b6), this state of affairs is not possible. The alternative, Aristotle reasons, is to have an everlasting cycle of generation that guarantees an endless supply of generated beings: 'For that way being would be most connected (*malista suneiroito*) because coming to be eternally is the closest thing generation has to being.' If the generation of sublunary beings were intermittent rather than continuous, so that there were periods in the earth's history when no sublunary beings existed, that would decrease the overall amount of being in the world. The world would contain existential gaps, as it were.[27] Perpetual generation avoids this by linking individuals together in a continuous (gapless) chain, so that, although each individual must pass out of existence, there will always be some other sublunary being to take its

[25] Aristotle treats this as an empirical truth based on observations about the way living things behave (e.g. *De juv.* 469a28–30).

[26] I see this kind of optimality reasoning, which explains why *S* is *P* by showing that *P* is the best way for *S* to be, as a sub-species of teleological explanation even though it does not involve the typical "for the sake of" (*hou heneka*) language. Compare *Phaedo* 97b8–98a2, where Socrates appeals to what is best as a cause (*aitia*) of things in nature. For a discussion of this type of explanation in Plato and Aristotle see Henry (2013). For a good example of this optimality reading see *IA* 708a9–20 on why snakes lack feet.

[27] This way of expressing the point was suggested to me by Christopher Frey during my session at the Princeton Classical Philosophy Colloquium (7 December 2013).

place.[28] *GC* II 10 makes this point in relation to the elements, and so we can begin there.

In this book I have argued that no *specific* instance of elemental transformation (which is a rectilinear process) occurs for the sake of the end that results. For example, when air comes to be from fire, fire is the subject from which the change proceeds, dry is the privation that is lost in the change, and moist is the form that is acquired as a result of the change. But there is no goal-directed efficient cause that directs the process towards that form (the moist is not the *telos* of the change). But there is a sense in which *the whole continuous cycle* of elemental transformations (and the sorts of processes that follow upon them, e.g. the rain cycle) exhibit beauty and goodness insofar as it is orderly and not haphazard. This is borne out by the *GC* II 10 passage. In *Physics* VIII 8 Aristotle had argued that only circular motion can be continuous, and hence everlasting, because unlike rectilinear motions its starting and end points are one and the same. This is the kind of motion exhibited by the celestial bodies. And since these do not come to be or pass away, it is always numerically the same substance that remains forever revolving in a circle. But elemental transformation (which is itself a rectilinear process) can achieve this necessary continuity by 'imitating' the circular motion of the celestial bodies. When water is transformed into air, one element (air) comes into being while another (water) passes away (*GC* 319b17–22). However, linking the passing-away of one element to the coming-to-be of another in a continuous cycle of reciprocal transformations avoids any existential gaps in this part of the natural world:

> That, too, is why all other things – the things, I mean, which are reciprocally transformed in virtue of their qualities and their powers, e.g. the simple bodies – imitate circular motion. For when water is transformed into air, air into fire, and then fire back into water we say that coming-to-be has completed the cycle because it reverts again to the beginning. Hence it is by imitating circular motion that rectilinear motion too is continuous. (337a1–7)

It is in this way that the being of the elements is most connected. Thus, it turns out that elemental transformation is the best possible way for the god to 'fill up' (*suneplêrôse*) the universe with this type of sublunary being.

What I want to suggest is that *GA* II 1, 731b24–732a3 is meant to apply this same general reasoning to the specific case of biological generation as

[28] As mentioned, this way of reading the argument does not commit Aristotle to the claim that the universe must somehow be benefited by continuous generation (see note 20 above).

part of the argument for why sexes exist. The *GA* argument at least appears to follow the same pattern. Being is better than non-being, living (which is being for an organism) is better than not living, and so forth. Sublunary organisms are incapable of eternal life since they contain matter organisms and so must pass out of existence. Thus, since there cannot be any eternal organisms in the sublunary world, the next best thing is to make their generation continuous. For having eternal lineages (a continuous generation of things of the same form) is the best possible way to maximise the amount of life in the world given the constraints imposed on living things by their material natures (and nature always strives after what is better given the range of possibilities). Eternal lineages thus ensure that biological existence is most connected by leaving no existential gaps in the living world. This, in turn, provides the final cause for the existence of sexes: males and females are present for the sake of keeping these lineages in existence insofar as they constitute the first principles of generation.

To close this section, let me offer some arguments for thinking that the *GA* 731b24–735a3 (translated above) is best understood as a specific application of the *GC* II 10 argument rather than simply treating it as a version of the *DA* II 4 argument.

The main reason scholars have tended to read the *GA* and *DA* passages together is that the thought expressed in text [B] is also part of the *DA* argument.[29] But [B] is also a premise in the *GC* II 10 argument. And so this alone cannot decide between the two readings. By contrast, the *GC* and *GA* arguments both draw on the same axiological considerations expressed in [A], which do not figure into the *DA* argument. While it may be conceded that these same principles are in the background of the *DA* argument, the fact that Aristotle makes explicit use of them in the *GA* passage provides a more explicit connection with the *GC* II 10 passage.

Furthermore, the *DA* and *GA* arguments do not even appear to be focused on the same feature of generation. In the *DA* argument Aristotle is attempting to explain the biological imperative to reproduce (415a26–b7), which he relates to a desire for personal immortality. The *GA* argument is an attempt to explain why sexes exist. It does that without explicitly relying on the idea that individuals engage in reproduction in order to 'partake in the everlasting and divine'. Instead, Aristotle explains the

[29] Text [B] states: 'For, since it is not possible for the nature of this kind of thing to be eternal, what comes into being is eternal in the only way possible for it. While it is not possible to be eternal *in number* (for the substance of existing things is in the particular; if it was of this sort, then it would be eternal), it is possible to be so *in form*. For this reason, there is always a continuous generation of humans, animals, and plants.'

existence of sexes by appealing to the fact that sexual reproduction is an eternal feature of the sublunary world.[30] This is precisely what the *GC* II 10 argument attempts to explain (at a more general level) by appealing to the very same considerations found in texts [A] and [B] of the *GA* argument.[31]

Finally, the explicit aim of the *DA* argument is to account for the fact that all organisms 'that are complete and not maimed or spontaneously generated' do everything they do in order to accomplish the natural function of reproduction ('to produce another like itself'), which Aristotle explains in terms of the individual's desire to partake in the everlasting and divine. This is Aristotle's only aim in that passage, namely, to explain the biological imperative to reproduce. The argument is not concerned to explain why there is always a *genos* of humans, animals, plants.

Nor does the *DA* argument appear to support that conclusion. I follow Lennox in taking *genos* here to have the sense given at *Metaphysics* 1024a29–30 ('the continuous generation of things having the same form'). To say there is always (*aei*) a continuous generation of these biological kinds is to say that their generation is perpetual and everlasting, which means that they will not only continue indefinitely into the future but that they also stretch indefinitely back into the past (they are unending and without beginning). Now the argument in texts [A] and [B] of the *GA* passage is supposed to provide the grounds for this key conclusion ('*For this reason*, there is always a *genos* of humans, animals, and plants'), which in turn supplies the final cause for the existence of the sexes.[32] By contrast, the *DA* argument does not support that conclusion and so it leaves a crucial part of the *GA*'s answer to why sexes exist unexplained.

First of all, the *DA* argument is not adequate to explain why sexual reproduction is unending (why there always *will be* a continuous generation of biological kinds). At best the desire for immortality can ensure that living things will leave future generations *if* nothing prevents them from successfully reproducing. However, Aristotle's doctrine about the eternity of species involves the much stronger claim that *necessarily* these ancestor–

30 This is a fairly standard way of reading the *GA* II 1 argument. For example, Cooper (1982, 203 n. 4) writes: 'Aristotle appeals to the permanence of animal and plant life to explain why there is sexual differentiation and animal reproduction in general' and that 'it is in order to sustain these genera in existence that reproduction through sexual differentiation takes place'. Even Gotthelf (2012, 58), who defends the organism-centred reading, takes the passage to be focused on 'explaining the unendingness of sexual generation'.

31 Cooper (1982, 202–5). 32 Compare *GA* IV 3, 767b8–10.

descendant lineages will continue on without fail. But suppose we accept that most organisms possess a deep-seated desire for personal immortality that drives them to reproduce. Further suppose that subsequent generations inherit this same desire that drives them to reproduce as well. If this continues indefinitely (and necessarily!), then this might be enough to warrant the conclusion that all current species will continue to reproduce indefinitely into the future. But the desire for immortality alone is not enough to explain why these ancestor–descendant lineages extend indefinitely into the past (i.e. that there *has always been* a continuous generation of these same biological kinds). The organism's desire for personal immortality is satisfied by ensuring that all of its descendants inherit its form (that what *remains* is something like it in form). But it makes no difference from the perspective of one's own personal immortality whether its ancestors share in that form, or whether it even has ancestors at all. At least Aristotle never suggests in *DA* II 4 that one's desire for personal immortality is equally fulfilled by being the end of an ancestor–descendant lineage that stretches eternally into the past. The *DA* argument is compatible with the view that sexual reproduction evolved at some point in the history of life and has been ongoing ever since.[33]

If this is right, then the *DA* argument does not explain (or even attempt to explain) the important conclusion at *GA* 731b35–6 that 'there is always a *genos* of humans, animals, and plants'. But that conclusion is supported by the *GC* II 10 argument, which says that making coming-to-be perpetual and everlasting contributes to the good condition of the universe by maximising the amount of sublunary being in the world. For this reason,

[33] Johansen (2015, 126–7) argues that the reference to two senses of end at *DA* 415b2–3 invites a distinction between the goal of sexual reproduction (the end "for the sake of which it" occurs) and its beneficiary (that "for the benefit of which" or "in whose interest" it occurs). He contends that, while the goal of formal replication is personal immortality for the individual, the beneficiary must be the species itself and not the individual. For the individual is not around to enjoy his own afterlife. In that case one might argue that even if the history of the species form is inconsequential from the perspective of *my personal* immortality (*I* am indifferent to whether my species form has existed before me; I only care about its continued existence), it is in the interest of the species itself. For the species benefits from being eternal in both directions. But this inference is not supported by the text. The *DA* II 4 argument claims that animals engage in reproduction for the sake of participating in the eternal. Even if we agree that something must benefit from the continued existence of its form (the *DA* II 4 argument makes no such claim), Johansen is too quick to assume that Aristotle shares his intuition that it cannot be the individual because it is not around to enjoy it. First, it is not clear that Aristotle would agree that one can only benefit from (or be harmed by) something if one is around to enjoy (or suffer) it (cf. *NE* I 10–11). Second, I find it plausible to think that Aristotle would share the modern thought that parents have children at least in part because they have an interest in the continued existence of their own genes, even though they know they will not be around to enjoy that good.

I am inclined to think that [A] and [B], which rely on the same axiological considerations as *GC* II 10 (e.g. being is better than non-being), are instead meant to be a reconstruction of that cosmological argument.

In this chapter I have offered two interpretations of the *GA* II 1 argument, which have very different implications for how Aristotle understood the teleological significance of biological generation itself. According to the organism-centred interpretation, the *GA* passage is merely echoing the point of the *DA* argument that reproduction exists because it provides *the individual* with a means of partaking in eternal being. The existence of eternal lineages (species) is then treated as an artefact of particular organisms pursuing their own good. By contrast, I take the argument in *GA* II 1 to be focused on the existence of eternal lineages themselves. On this reading, eternal lineages are treated not as a mere consequence of individuals striving after personal immortality but as the best possible means of 'filling up' the universe with being given what is possible for a living thing. In this context, continuous generation is explained (following the *GC* II 10 argument) by the contribution it makes to the best possible state of the universe as a whole. The existence of the sexes is then explained as something that is conditionally necessary for that. This gives the sexes more cosmological significance than the organism-centred view allows.[34]

9.4 Necessity and Eternity

To close this book I want to turn to *GC* II 11 and consider the relation between necessity and the continuity of biological generation. Aristotle's discussion in this chapter is extremely challenging, and its arguments are not easy to decipher. My aim here is not to provide a decisive reading but simply to motivate a particular interpretation.

At the outset of the chapter Aristotle says that every continuous process of change involves consecutiveness where one thing comes to be after another with no interval between them (337a35–b2). The aim of the chapter is to investigate, in the case of coming to be, whether any members of the series are absolutely necessary or whether they are all merely contingent in the sense that they could fail to come into being:

[34] The above argument should not be taken to vindicate the kind of Cosmic Teleology reading that Sedley endorses. In Henry (2015b, 112–14) I offer several reasons why I think Sedley goes too far in claiming that Aristotle thinks of the "parts" of the universe as somehow being adjusted to one another in such a way that their mutual interactions contribute to the cosmic good and, ultimately, the good of man.

> Then question is whether all <generated> things are of this sort or whether for some of them it is necessary in the strict sense (*anankaion haplôs*) that they come into being. In other words, is it the case that, just as in the sphere of being some things cannot possibly fail to exist, so too concerning generation [sc. some things cannot possibly fail to come into being]? For example, is it necessary that the solstices will occur in the sense that it is not possible for them to be incapable of occurring? (337b10–13)

Joachim calls *GC* II 11 an 'appendix' because he thinks the main task of *GC* is completed in *GC* II 10 and that the discussion of necessity in the final chapter adds nothing to this. But the discussion of necessity is clearly connected to *GC* II 10 and the theme of continuous generation. As I read the chapter, Aristotle is ultimately worried about the possibility that generation may not continue eternally but could cease at some point. Aristotle will go on to argue that in every process of generation it is never necessary in the strict sense that the thing generated will come to be; its generation is always contingent. This is especially true in the case of living things, where the generation of each individual is dependent on a number of complex factors that are themselves contingent (e.g. one or both of its parents might have been infertile, or they might have failed to conceive, or its development might have been derailed due to some accident in the process, and so forth). But if it is possible that each individual might fail to come into being, then what is the source of that necessity that is supposed to make biological generation itself eternal? What ensures that there will always be a *genos* of humans, animals, and plants?[35]

The argument of *GC* II 11 proceeds in three parts. (1) Aristotle first presents a series of arguments to show that it is not necessary in the strict sense for any generated substance to come into being; each one is such that it could have failed to come to be (337b14–33). Its coming-to-be is only 'conditionally' necessary, as Aristotle puts it.[36] From this he concludes that what is necessary in the absolute sense (*haplôs*) must occur eternally and vice versa (337b33–338a3; cf. *NE* VI 3, 1139b20–4).[37] (2) Next Aristotle argues that it is in circular motion, and in cyclical generation in particular,

[35] *GC* II 11 starts off as a fairly general worry about the necessity of all generated substances, but towards the end of the chapter it becomes clear that Aristotle's main concern is the eternity of biological generation (see 338b6–11 and note 39).

[36] *GC* 337b14–33 provides a set of complicated arguments to show that, in any causal sequence where some prior member *A* is the cause of some subsequent member *B*, both *A* and *B* will always *only* be conditionally necessary. Joachim (1926, 272–4) offers a good exposition of those arguments. I reserve judgement here on whether or not this is the same conditional necessity that is operative in the biological works (e.g. *PA* I 1, 639b21–30; cf. Cooper 1982).

[37] Compare Charles (1988, 14–17).

that such absolute (*haplôs*) necessity is to be found (338a3–18).[38] (3) Finally, he raises a worry about the generation of living things:

> But then why do some things obviously come to be in this way (as, for instance, rain and air come to be cyclically so that it must rain if there is a cloud and, conversely there must be a cloud if it is to rain), while humans and animals do not return upon themselves so that the same individual comes to be for a second time (for although your coming to be presupposes your father's, his coming to be does not presuppose yours)? Why, on the contrary, does this sort of generation seem to proceed in a straight line? (338b6–11)

Back in *GC* II 10 Aristotle had argued that only continuous motion is properly eternal and that circular motion is the only kind of motion that is genuinely continuous. However, a rectilinear process, like natural generation, can 'imitate' circular motion by reverting back on itself thereby forming a cycle (think of the rain cycle where the elements are reciprocally transformed into one another). Here Aristotle is worried that the generation of living things is not cyclical in the right way. This is a problem because he has just argued that what is necessary *haplôs* is eternal and that it is only in cyclical generation that this sort of necessity is to be found. What Aristotle needs to show, then, is that the generation of living things is necessary in the strict sense while respecting the fact that each individual organism is only conditionally necessary and could fail to come into being (cf. 337b26–7). In a word, he needs to show how *eternal* lineages can be composed of a series of *contingent* beings.[39] Let me flesh this out a bit, since this is not the traditional way of reading the chapter.

[38] Joachim (1926, 273) summarises the doctrine thus: 'No member of a rectilinear succession of *gignomena* ... can exhibit "absolute necessity of occurrence". If a thing is to come-to-be with "absolute necessity", it must come-to-be always and invariably: and that is possible only if it is a member of an eternally-repeated cycle of *gignomena* (37b33–38a5). Hence "absolute necessity of occurrence" and "reciprocal necessary *nexus*" (which depends on it) are to be found only in cyclical *kinêsis* and cyclical *genesis* (38a5–17).'

[39] My interpretation, which takes Aristotle to be worried primarily about the fate of biological generation, is suggested by the puzzle at 338b6–11 (towards which the entire chapter appears to have been building) where the relevant contrast is between meteorological phenomena and heavenly motions, on the one hand, and biological generation, on the other. There are two reasons for thinking Aristotle is not concerned with the continuity (and therefore eternity) of elemental transformation in *GC* II 11. First, Aristotle has already laid it down in *GC* II 10 that elemental transformation is a continuous process of generation in which the elements "return upon" themselves and thereby "imitate" circular motion (337a1–7). Second, the *GC* II 10 passage appears to have the rain cycle in mind, which is one of those processes of generation that 338b6–11 takes to be 'obviously' continuous/cyclical. For this reason, I take *GC* II 11 to be raising a new worry about the continuity (and thus eternity) of biological generation. While it is true that Aristotle mentions the elements in the final paragraph of the chapter, he does so only to illustrate the kind of solution he wants to apply to the case of living things. The point is to extend what he said about the elements in

It is often supposed that Aristotle thinks biological generation is only necessary in the conditional sense. Leunissen, for example, argues that absolute necessity holds for eternal things such as the movement of the heavens while 'conditional necessity holds for the generation of animals, which is a sublunary natural process that is rectilinear and that concerns beings whose substance is perishable'.[40] Leunissen seems to take Aristotle's statement about biological generation at 338b6–11 to express his settled position in *GC* II 11. But, first, we should note that Aristotle only says the process seems to be (*eoiken*) rectilinear (338b11). I take this to be a tentative statement that is part of the set up for the problem in question. Second, it is important to distinguish between the claim that the generation of some particular organism (e.g. Socrates) is only conditionally necessary from the claim that the phenomenon of biological generation itself is only conditionally necessary. Leunissen is right that Aristotle thinks absolute necessity holds only for eternal things, while the generation of *each particular* organism is a rectilinear process that involves only conditional necessity. He seems to say as much at 338b6–11 (cf. 337a25–9). But what Aristotle wants to say (and, in fact, goes on to say, if I am reading the text properly) is that *biological generation itself* is a cyclical process and therefore necessary in the strict sense. As we have seen, the *GA* II 1 passage contains an argument for why the generation of living things will continue forever so that there will always be a *genos* of humans, animals, and plants (731b35–6). And, according to *GC* II 11, what occurs always is necessary in the unqualified sense. Herein lies the paradox. Aristotle thinks that the coming-to-be of each particular organism is not absolutely necessary but only conditionally necessary so that each individual could have failed to come into being. But if so, then it should be the case that biological generation itself could cease at some point.

We have seen that Aristotle thinks only continuous motion is properly eternal and that only circular motion is genuinely continuous. But he also holds that rectilinear motion can imitate circular motion by being cyclical (i.e. reverting back on itself). And it is in cyclical generation, he says, that absolute necessity is to be found. Aristotle says that, at first glance, it seems as though (*eoiken*) biological generation is not cyclical in the right way; for 'animals do not return upon themselves so that *the same individual* comes to be a second time' (338b8–9). Once a particular animal passes out of

GC II 10 to this central case (living things must, like the elements, return upon themselves, not numerically, but formally).

[40] Leunissen (2010, 105).

existence, it is gone forever. By contrast, it is numerically the same sun that returns again and again to the same position every time it revolves around the earth. This is where I think many commentators leave the argument. But Aristotle continues by distinguishing between two different ways in which a rectilinear causal process might be cyclical and so imitate circular motion:

> In discussing this, we must begin by inquiring whether all things return upon themselves in a uniform manner or whether, on the contrary, though in some sequences what recurs is *numerically* the same, in other sequences it is only *the same in form*. Of those moving substances that are imperishable, it is clear that <what returns upon itself> will be numerically the same (for the motion is consequent upon that which is being moved). But those things that are perishable must return upon themselves formally but not numerically. This is why, when water comes to be from air and air from water, the air <that returns> is the same in form not the same in number. And even if these are numerically the same, this, at any rate, is not the case with those things whose substance comes into being, being the sort of thing that is capable of being otherwise. (338b6–19)

Aristotle thinks biological generation is an instance of this kind of formal replication where what returns is not numerically the same individual but something that is the same in form. In this way reproduction itself takes on the character of a cyclical process. And since all cyclical processes are necessary in the strict sense, and since what is absolutely necessary is eternal, it follows that the substantial generation must be eternal and continue without fail.

Bibliography

Ainsworth, T. 2016. 'Form vs. matter', in E. N. Zalta (ed.), *The Stanford Encyclopedia of Philosophy* (Spring 2016), http://plato.stanford.edu/archives/spr2016/entries/form-matter/.

Algra, K. 2004. 'On *Generation and Corruption* I. 3: substantial change and the problem of not-being', in F. de Haas and J. Mansfeld, *Aristotle: On Generation and Corruption, Book I*. Symposium Aristotelicum, pp. 91–121. Oxford: Clarendon Press.

Anagnostopoulos, A. 2013. 'Aristotle's Parmenidean dilemma', *Archiv für Geschichte der Philosophie* 95(3): 245–74.

Annas, J. 1982. 'Aristotle on inefficient causes', *The Philosophical Quarterly* 32 (129): 311–26.

Balme, D. [1972](1992). *Aristotle: De partibus animalium I and De generatione animalium I (with passages from II.1–13)*. Translated with notes, with a report on recent work and additional bibliography by A. Gotthelf. Oxford: Clarendon Press.

1987. 'Aristotle's biology was not essentialist', in A. Gotthelf and J. G. Lennox (eds.), *Philosophical Issues in Aristotle's Biology*, pp. 291–312. Cambridge: Cambridge University Press.

1962. 'GENOS and EIDOS in Aristotle's biology', *The Classical Quarterly*, New Series 12(1): 81–98.

Barnes, J. 2000. *Aristotle: A Very Short Introduction*. Oxford: Oxford University Press.

1982. *Aristotle*. Oxford: Oxford University Press.

Beere, J. 2009. *Doing and Being: An Interpretation of Aristotle's Metaphysics Theta*. Oxford: Oxford University Press.

Bodnar, I. 2012. 'Aristotle's natural philosophy', in E. N. Zalta (ed.), *The Stanford Encyclopedia of Philosophy* (Spring 2012), http://plato.stanford.edu/archives/spr2012/entries/aristotle-natphil/.

2005. 'Teleology across nature', *Rhizai* 2: 9–29.

1997. 'Movers and elemental motion in Aristotle', *Oxford Studies in Ancient Philosophy* 15: 81–117.

Bolton, R. 2015. 'Aristotle's natural teleology in *Physics* II', in M. Leunissen (ed.), *Aristotle's Physics: A Critical Guide*, pp. 121–43. Cambridge: Cambridge University Press.

2010. 'Biology and metaphysics in Aristotle', in J. G. Lennox and R. Bolton (eds.), *Being, Nature, and Life in Aristotle*, pp. 30–55. Cambridge: Cambridge University Press.

1987. 'Definition and scientific method in Aristotle's *Posterior Analytics* and *Generation of Animals*', in A. Gotthelf and J. G. Lennox (eds.), *Philosophical Issues in Aristotle's Biology*, pp. 120–66. Cambridge: Cambridge University Press.

Bostock, D. 2006. *Space, Time, Matter, and Form: Essays on Aristotle's Physics*. Oxford: Clarendon Press.

1994. *Aristotle: Metaphysics Books Z and H*. Translation with commentary. Oxford: Clarendon Press.

Broadie, S. 2004. 'On *Generation and Corruption* I. 4: distinguishing alteration', in F. de Haas and J. Mansfeld (eds.), *Aristotle: On Generation and Corruption, Book I*. Symposium Aristotelicum, pp. 123–50. Oxford: Clarendon Press.

Broadie [Waterlow], S. 1982. *Nature, Change, and Agency in Aristotle's Physics*. Oxford: Clarendon Press.

Brown, L. 1994. 'The verb "to be" in Greek philosophy', in S. Everson (ed.), *Language (Companions to Ancient Thought 3)*, pp. 212–36. Cambridge: Cambridge University Press.

1986. 'Being in the Sophist: a syntactic inquiry', *Oxford Studies in Ancient Philosophy* 4: 49–70. Reprinted in G. Fine (ed.), *Plato 1: Metaphysics and Epistemology*, 1999, pp. 455–78, Oxford: Oxford University Press.

Brunschwig, J. 2004. 'On *Generation and Corruption* I. 1: a false start?', in F. de Haas and J. Mansfeld (eds.), *Aristotle: On Generation and Corruption, Book I*. Symposium Aristotelicum, pp. 25–63. Oxford: Clarendon Press.

Burnyeat, M. 2004. 'Aristotle on the foundations of sublunary physics', in F. de Haas and J. Mansfeld (eds.), *Aristotle:* On Generation and Corruption, *Book I*. Symposium Aristotelicum, pp. 7–24. Oxford: Clarendon Press.

Carraro, N. 2017. 'Aristotle's embryology and Ackrill's Problem', *Phronesis* 62(3): 274–304.

Charles, D. 2018. '*Physics* I 7', in D. Quarantotto (ed.), *Aristotle's Physics Book I: A Systematic Exploration*, pp. 178–205. Cambridge: Cambridge University Press.

2010. *Definition in Greek Philosophy*. Oxford: Oxford University Press.

2004. 'Simple genesis and prime matter', in F. de Haas and J. Mansfeld (eds.), *Aristotle: On Generation and Corruption, Book I*. Symposium Aristotelicum, pp. 151–69. Oxford: Clarendon Press.

2001. *Aristotle on Meaning and Essence*. Oxford: Oxford University Press.

1991. 'Teleological causation in the *Physics*', in L. Judson (ed.), *Aristotle's Physics: A Collection of Essays*, pp. 101–28. Oxford: Oxford University Press.

1988. 'Aristotle on hypothetical necessity and irreducibility', *Pacific Philosophical Quarterly* 69: 1–53.

Charlton, W. 1970. *Aristotle's Physics Books 1 & 2*. Translation with commentary. Oxford: Clarendon Press.

Code, A. 2003. 'Changes, powers and potentialities in Aristotle', in N. Reshotko (ed.), *Desire, Identity and Existence: Essays in Honor of T. M. Penner*, pp. 251–71. Kelowna: Academic Printing & Publishing.

1995. 'Potentiality in Aristotle's Science and Metaphysics', *Pacific Philosophical Quarterly* 76(3/4): 405–18.

1987. 'Soul as efficient cause in Aristotle's embryology'. *Philosophical Topics* 15 (2): 51–9.

1976. 'The persistence of Aristotelian matter', *Philosophical Studies* 29(6): 357–67.

Connell, S. 2016. *Aristotle on Female Animals*. Cambridge: Cambridge University Press.

2001. 'An integrated approach to Aristotle as a biological philosopher', *Review of Metaphysics* 55(2): 297–322.

Cook, K. 1996. 'Sexual inequality in Aristotle's theories of reproduction and inheritance' in Julie K. Ward (ed.), *Feminism and Ancient Philosophy*, pp. 51–67. New York: Routledge.

Cooper, J. 1988. 'Metaphysics in Aristotle's embryology', *Cambridge Philological Society Proceedings*, 214 (new series 34): 14–41.

1987. 'Hypothetical necessity and natural teleology', in A. Gotthelf and J. G. Lennox (eds.), *Philosophical Issues in Aristotle's Biology*, pp. 243–74. Cambridge: Cambridge University Press.

1982. 'Aristotle on natural teleology', in M. Schofield and M. C. Nussbaum (eds.), *Language and Logos*, pp. 197–222. Cambridge: Cambridge University Press.

Corcilius, K. and P. Gregoric. 2010. 'Separability vs. difference: parts and capacities of the soul in Aristotle', *Oxford Studies in Ancient Philosophy* 39: 81–120.

Curd, P. 2011. *A Presocratics Reader: Selected Fragments and Testimonia*. Indianapolis: Hackett Publishing.

De Haas, F. and J. Mansfeld, eds. 2004. *Aristotle: On Generation and Corruption, Book I*. Symposium Aristotelicum. Oxford: Clarendon Press.

Ebrey, D. In press. 'The matter for birth, life, and the elements', in L. Taub (ed.), *Cambridge Companion to Ancient Science*. Cambridge: Cambridge University Press.

2018. 'The definition of matter in *Physics* I', unpublished paper.

2007. 'Aristotle's motivation for matter', unpublished PhD dissertation, University of California, Berkeley.

Ebrey, D. (ed.) 2015. *Theory and Practice in Aristotle's Natural Science*. Cambridge: Cambridge University Press.

Falcon, A. 2005. *Aristotle and the Science of Nature*. Cambridge: Cambridge University Press.

Falcon, A. and M. Leunissen 2015. '*Eulogos* in natural scientific inquiry', in D. Ebrey, *Theory and Practice in Aristotle's Natural Philosophy*, pp. 217–40. Cambridge: Cambridge University Press.

Fankhauser, D. 2017. 'Rennet for making cheese', www.biology.clc.uc.edu/fankhauser/Cheese/Rennet/Rennet.html.

Ferejohn, M. 1989. 'Plato and Aristotle on negative predication', *Archiv für Geschichte der Philosophie* 71(3): 257–82.

Frede, D. 2004. '*On Generation and Corruption* I. 10: on mixtures and mixables', in F. de Haas and J. Mansfeld, *Aristotle: On Generation and Corruption, Book I.* Symposium Aristotelicum, pp. 289–314. Oxford: Clarendon Press.

Frede, M. 1994. 'Aristotle's notion of potentiality in *Metaphysics*', in T. Scaltas, D. Charles, and M.-L. Gill (eds.), *Unity, Identity, and Explanation in Aristotle's Metaphysics*, pp. 173–94. Oxford: Clarendon Press.

Freeland, C. 1987. 'Aristotle on bodies, matter, and potentiality', in A. Gotthelf and J. G. Lennox (eds.), *Philosophical Issues in Aristotle's Biology*, pp. 392–407. Cambridge: Cambridge University Press.

Freudenthal, G. 1999. *Aristotle's Theory of Material Substance: Heat and Pneuma, Form and Soul.* Oxford: Clarendon Press.

Frey, C. 2015. 'From blood to flesh: homonymy, unity, and ways of being in Aristotle', *Ancient Philosophy* 35(2): 375–94.

2007. 'Organic unity and the matter of man', *Oxford Studies in Ancient Philosophy* 32: 167–204.

Furth, M. 1988. *Substance, Form, and Psyche: An Aristotelian Metaphysics.* Cambridge: Cambridge University Press.

Gelber, J. 2010. 'Form and inheritance in Aristotle's embryology', *Oxford Studies in Ancient Philosophy* 39: 183–212.

Gill, M-L. 2016. 'Aristotle's hylomorphism in *Metaphysics* Theta', unpublished paper.

2014. 'The limits of teleology in Aristotle's *Meteorology* IV.12', *HOPOS: The Journal of the International Society for the History of Philosophy of Science* 4: 335–50.

2012. *Philosophos: Plato's Missing Dialogue.* Oxford: Oxford University Press.

1989. *Aristotle on Substance: The Paradox of Unity.* Princeton: Princeton University Press.

Gotthelf, A. 2012. *Teleology, First Principles, and Scientific Method in Aristotle's Biology.* Oxford: Oxford University Press.

Gotthelf, A. and J. G. Lennox, eds. 1987. *Philosophical Issues in Aristotle's Biology.* Cambridge: Cambridge University Press.

Henry, D. 2018. 'Aristotle on animals', in P. Adamson and G. F. Edwards (eds.), *Animals: A History*, pp. 9–26. New York: Oxford University Press.

2017. 'Aristotle on epigenesis', in A. Falcon and D. Lefebvre (eds.), *Aristotle's Generation of Animals: A Critical Guide*, pp. 89–107. Cambridge: Cambridge University Press.

2015a. 'Substantial generation in *Physics* I.5–7', in M. Leunissen (ed.), *Aristotle's Physics: A Critical Guide*, pp. 144–61. Cambridge: Cambridge University Press.

2015b. 'The cosmological significance of animal generation', in D. Ebrey (ed.), *Theory and Practice in Aristotle's Natural Science*, pp. 100–18. Cambridge: Cambridge University Press.

2014. 'The birds and the bees: Aristotle on the biological concept of analogy', in G. M. Gurtler SJ and W. Wians (eds.), *Proceedings of the Boston Area Colloquium in Ancient Philosophy*, pp. 145–69. Boston: Brill.

2013. 'Optimality reasoning in Aristotle's natural teleology', *Oxford Studies in Ancient Philosophy* 45: 225–63.

2011. 'Review of Explanation and Teleology in Aristotle's Science of Nature', *Notre Dame Philosophical Reviews* 11 May, http://ndpr.nd.edu/news/explanation-and-teleology-in-aristotle-s-science-of-nature/.

2009. 'Aristotle's *Generation of Animals*', in G. Anagnostopoulos (ed.), *A Companion to Aristotle*, pp. 368–83. Malden: Wiley-Blackwell.

2008. 'Organismal natures', in J. Mouracade (ed.), *Aristotle on Life*. Special issue of *Apeiron: A Journal for Ancient Philosophy and Science* 41(3): 47–74.

2007a. 'How sexist is Aristotle's developmental biology?', *Phronesis* 52(3): 251–69.

2007b. 'Review of Monte Ransome Johnson *Aristotle on Teleology*', *Ancient Philosophy* 27(1): 191–200.

2006a. 'Aristotle on the mechanisms of inheritance', *Journal of the History of Biology* 39: 425–55.

2006b. 'Understanding Aristotle's reproductive hylomorphism', *Apeiron: A Journal for Ancient Philosophy and Science* 39(3): 257–88.

2005. 'Embryological models in ancient philosophy', *Phronesis* 50(1): 1–42.

Irwin, T. H. 2010. 'The sense and reference of *Kalon* in Aristotle', *Classical Philology* 105(4): 381–96.

Joachim, H. H. 1951. *Aristotle Nicomachean Ethics*. Translation with commentary. Oxford: Clarendon Press.

1926. *Aristotle on Coming-to-Be and Passing-Away*. Revised text with introduction and commentary. Oxford: Clarendon Press.

Johansen, T. K. 2015. 'The two kinds of ends in Aristotle: the view from *de Anima*', in D. Ebrey (ed.), *Theory and Practice in Aristotle's Natural Science*, pp. 119–36. Cambridge: Cambridge University Press.

2012. *The Powers of Aristotle's Soul*. Oxford: Oxford University Press.

2004. *Plato's Natural Philosophy: A Study of the Timaeus–Critias*. Cambridge: Cambridge University Press.

Johnson, M. R. 2015. 'Aristotle's architectonic sciences', in D. Ebrey (ed.), *Theory and Practice in Aristotle's Natural Science*, pp. 163–86. Cambridge: Cambridge University Press.

2005. *Aristotle on Teleology*. Oxford: Oxford University Press.

Jones, B. 1974. 'Aristotle's Introduction of Matter', *The Philosophical Review* 83(4): 474–500.

Judson, L. 2005. 'Aristotelian teleology', *Oxford Studies in Ancient Philosophy* 29: 341–66.

Kahn, C. 2009. *Essays on Being*. Oxford: Oxford University Press.

1981. 'Some philosophical uses of "to be" in Plato', *Phronesis* 26(2): 105–34.

1986. 'Retrospective on the verb "to be" and the concept of being', in S. Knuuttila and J. Hintikka (eds.), *The Logic of Being*, pp. 1–28. Dordrecht: Reidel.

Katz, E. 2017. 'Commentary on *GC* II 9', paper presented at the *Notre Dame Centre for Aristotelian Studies Workshop*, University of Notre Dame, 13 May 2017.

Kelsey, S. 2015. 'Aristotle on interpreting nature', in M. Leunissen (ed.), *Aristotle's Physics: A Critical Guide*, pp. 31–45. Cambridge: Cambridge University Press.

2010. 'Hylomorphism in Aristotle's *Physics*', *Ancient Philosophy* 30(1): 107–24.

2008. 'The place of I 7 in the argument of *Physics* I', *Phronesis* 53(2): 180–208.

2006. 'Aristotle *Physics* I 8', *Phronesis* 51(4): 330–61.

2003. 'Aristotle's definition of nature', *Oxford Studies in Ancient Philosophy* 25: 59–87.

Kono, T., Y. Obata, Q. Wu, K. Niwa, Y. Ono, Y. Yamamoto, E. S. Park, J. S. Seo, and H. Ogawa 2004. 'Birth of parthenogenetic mice that can develop to adulthood', *Nature* 428(6985): 860–4.

Kosman, L. A. 2010. 'Male and female in Aristotle's *Generation of Animals*', in James G. Lennox and Robert Bolton (eds.), *Being, Nature, and Life in Aristotle*, pp. 147–67. Cambridge: Cambridge University Press.

Krizan, M. 2013. 'Substantial change and the limiting case of Aristotelian matter', *History of Philosophy Quarterly* 30(4): 293–310.

Kullman, W. (1985). 'Different concepts of the final cause in Aristotle', in A. Gotthelf (ed.), *Aristotle on Nature and Living Things: Philosophical and Historical Studies Presented to David M. Balme on His Seventieth Birthday*, pp. 170–5. Pittsburgh: Mathesis Publications.

Kupreeva, I. 2010. 'Aristotle on causation and conditional necessity: *An. Post.* 2.12 in context', in F. de Haas, M. Leunissen, and M. Martijn (eds.), *Interpreting Aristotle's Posterior Analytics in Late Antiquity and Beyond*, pp. 203–34. Leiden: Brill.

Lane, M. S. 1998. *Method and Politics in Plato's Statesman*. Cambridge: Cambridge University Press.

Lang, H. 1998. *The Order of Nature in Aristotle's Physics*. Cambridge: Cambridge University Press.

Lear, J. 1988. *Aristotle: The Desire to Understand*. New York: Cambridge University Press.

Lefebvre, D. 2019. 'Looking for the formative power in Aristotle's nutritive soul', in R. Lo Presti and G. Korobili (eds.), *Nutrition and Nutritive Soul in Aristotle and Aristotelianism*. Berlin: De Gruyter.

Lennox, J. G. 2018. '"Most natural among the functions of living things": reproduction as a nutritive function', unpublished paper.

2010. '*Bios* and explanatory unity in Aristotle's biology', in D. Charles (ed.), *Definition in Greek Philosophy*, pp. 329–55. Oxford: Oxford University Press.

2008. '"As if we were investigating snubness": Aristotle on the prospects for a single science of nature', *Oxford Studies in Ancient Philosophy* 35: 149–86.

2004a. 'Aristotle on the emergence of material complexity: *Meteorologica* IV and Aristotle's biology', *HOPOS: The Journal of the International Society for the History of Philosophy of Science* 4(2): 272–305.

2004b. 'The place of zoology in Aristotle's natural philosophy', in R. W. Sharples (ed.), *Philosophy and the Sciences in Antiquity*, pp. 58–70. London: Ashgate.

2001a. *Aristotle's Philosophy of Biology*. Cambridge: Cambridge University Press.

2001b. 'Aristotle on the unity and disunity of science', *International Studies in the Philosophy of Science* 15(2): 133–44.

2001c. *Aristotle on the Parts of Animals I-IV*. Translation with introduction and commentary. Oxford: Clarendon Press.

1999. 'Teleology', in E. F. Keller and E. A. Lloyd (eds.), *Keywords in Evolutionary Biology*, pp. 324–23. Cambridge: Harvard University Press.

1994. 'Putting philosophy of science to the test: the case of Aristotle's biology', *PSA: Proceedings of the Biennial Meeting of the Philosophy of Science Association* 2: 239–47.

1985. 'Are Aristotelian Species Eternal?', in Allan Gotthelf (ed.), *Aristotle on Nature and Living Things*, pp. 67–94. Pittsburgh: Mathesis Publications.

1982. 'Teleology, chance, and Aristotle's theory of spontaneous generation', *Journal of the History of Philosophy* 20(3): 219–38.

Leunissen, M. 2012a. 'Crafting natures: Aristotle on animal design', in G. Dicker (ed.), *The Annual Proceedings of the Center for Philosophic Exchange*, pp. 28–51. Brockport: SUNY.

2012b. 'Aristotle on natural character and its implications for moral development', *Journal of the History of Philosophy* 50(4): 507–30.

2010. *Explanation and Teleology in Aristotle's Science of Nature*. Cambridge: Cambridge University Press.

Leunissen, M. and A. Gotthelf. 2010. 'What's teleology got to do with it? A reinterpretation of *Generation of Animals* V', *Phronesis* 55(4): 325–56.

Lewis, E. 2014. *Alexander of Aphrodisias: On Aristotle Meteorology 4*. Translation with introduction by E. Lewis. London: Duckworth.

1994. *Alexander of Aphrodisias: On Aristotle Meteorology 4*. London: Bloomsbury Publishing.

Lloyd, G. E. R. 1987. 'Empirical research in Aristotle's biology', in A. Gotthelf and J. G. Lennox (eds.), *Philosophical Issues in Aristotle's Biology*, pp. 53–63. Cambridge: Cambridge University Press.

Loux, M. J. 1992. 'Aristotle and Parmenides: an interpretation of *Physics* A8', in J. Cleary and W. Wians (eds.), *Proceedings of the Boston Area Colloquium in Ancient Philosophy*, pp. 281–319. Lanham: University Press of America.

Mason, A. 2006. 'Plato on necessity and chaos', *Philosophical Studies* 127(2): 283–98.

Matthews, G. 1982. 'Accidental unities', in M. Schofield and M. C. Nussbaum (eds.), *Language and Logos: Studies in Ancient Greek Philosophy*, pp. 223–40. Cambridge: Cambridge University Press.

Mayhew, R. 2004. *The Female in Aristotle's Biology*. Chicago: University of Chicago Press.

Menn, S. 2002. 'Aristotle's definition of soul and the programme of the *De Anima*', *Oxford Studies in Ancient Philosophy* 22: 83–139.

Meyer, S. Suavé 1992. 'Aristotle, teleology, and reduction', *Philosophical Review* 101(4): 791–825.

Mohr, R. 1982. 'The world-soul in the Platonic cosmology', *Illinois Classical Studies* 7(1): 41–8.

Morrow, G. R. 1950. 'Necessity and persuasion in Plato's *Timaeus*', *Philosophical Review* 59(2): 147–63.

Notomi, N. 1999. *The Unity of Plato's Sophist: Between the Sophist and the Philosopher*. Cambridge: Cambridge University Press.

Owen, G. E. L. 1999. 'Plato on not-being', in G. Fine (ed.), *Plato 1: Metaphysics and Epistemology*, pp. 416–54. Oxford: Oxford University Press. Originally published in G. Vlastos (ed.) 1970, *Plato, i: Metaphysics and Epistemology*, New York: Doubleday Anchor.

Palmer, J. 2008. 'Parmenides', in E. N. Zalta (ed.), *The Stanford Encyclopedia of Philosophy* (Feb. 2008) http://plato.stanford.edu/archives/fall2008/entries/parmenides/.

Peck, A. L. 1942. *Aristotle: Generation of Animals*. Translation with notes. Cambridge: Harvard University Press.

Pellegrin, P. 1986. *Aristotle's Classification of Animals: Biology and the Conceptual Unity of the Aristotelian Corpus*. Translated into English by A. Preus. Berkeley: University of California Press.

1985. 'Aristotle: zoology without species', translated into English by A. Preus, in A. Gotthelf (ed.), *Aristotle on Nature and Living Things*, pp. 95–116. Pittsburgh: Mathesis Publications.

Quarantotto, D. 2015. 'A dynamic ontology: on how Aristotle arrived at the conclusion that eternal change accomplishes *ousia*', in M. Leunissen (ed.), *Aristotle's Physics: A Critical Guide*, pp. 162–85. Cambridge: Cambridge University Press.

Robinson, H. M. 1974. 'Prime matter in Aristotle', *Phronesis* 19(2): 168–88.

Rosen, Jacob. 2015. '*Physics* V–VI versus VIII: unity of change and disunity in the *Physics*', in M. Leunissen (ed.), *Aristotle's Physics: A Critical Guide*, pp. 206–24. Cambridge: Cambridge University Press.

Ross, W. D. 1953. *Aristotle: Metaphysics*. Oxford: Clarendon Press.

1936. *Aristotle: Physics*. Translation with commentary. Oxford: Clarendon Press.

Scharle, M. 2015. '"And these things follow": teleology, necessity, and explanation in Aristotle's *Meteorologica*', in D. Ebrey (ed.), *Theory and Practice in Aristotle's Natural Philosophy*, pp. 79–99. Cambridge: Cambridge University Press.

2008. 'Elemental teleology in Aristotle's *Physics* 2.8', *Oxford Studies in Ancient Philosophy* 34: 147–83.

Schut, E., N. Hemmings, and T. R. Brickhead. 2008. 'Parthenogenesis in a passerine bird, the Zebra Finch *Taeniopygia guttata*', *Ibis* 150(1): 197–9.

Sedley, D. 2010. 'Teleology: Aristotelian and Platonic', in J. G. Lennox and R. Bolton (eds.), *Being, Nature, and Life in Aristotle*, pp. 5–29. Cambridge: Cambridge University Press.

2007. *Creationism and Its Critics in Antiquity*. Berkeley: University of California Press.

2004. 'On Generation and Corruption I. 2', in F. de Haas and J. Mansfeld (eds.), *Aristotle: On Generation and Corruption, Book I.* Symposium Aristotelicum, pp. 65–89. Oxford: Clarendon Press.

1991. 'Is Aristotle's teleology anthropocentric?', *Phronesis* 36(2): 179–96.

Segev, M. 2015. 'Review of Theory and Practice in Aristotle's Natural Science, by David Ebrey', *Notre Dame Philosophical Review* 2 November, https://ndpr.nd.edu/news/theory-and-practice-in-aristotle-s-natural-science.

Shields, C. 2016. *Aristotle: De anima.* Translation with commentary. Oxford: Clarendon Press.

2007. 'Substance and life in Aristotle', in J. Mouracade (ed.), *Aristotle on Life.* Special issue of *Apeiron: A Journal for Ancient Philosophy and Science*, 41(3): 129–52.

Sorabji, R. 1980. *Necessity, Cause, and Blame.* London: Duckworth.

Stavrianeas, S. 2015. 'Nature as a principle of change', in M. Leunissen (ed.), *Aristotle's Physics. A Critical Guide*, pp. 46–65. Cambridge: Cambridge University Press.

Strange, S. 1999. 'The double explanation in Plato's *Timaeus*', in G. Fine (ed.), *Plato 1. Metaphysics and Epistemology*, pp. 397–415. Oxford: Clarendon Press.

Tredennick, H. 1933. *Aristotle's Metaphysics.* Translated with introduction. Cambridge: Harvard University Press.

Tress, D. McGowan 1999. 'Aristotle against the Hippocratics on sexual generation: a reply to Coles', *Phronesis* 44(3): 228–41.

1992. 'The metaphysical science of Aristotle's *Generation of Animals* and its feminist critics', *The Review of Metaphysics* 46(2): 307–41.

Vlastos, G. 1969. 'Reasons and causes in the *Phaedo*', *Philosophical Review* 78(3): 291–325.

Wardy, R. 1993. 'Aristotelian rainfall and the lore of averages', *Phronesis* 38(1): 18–30.

Whiting, J. 2002. 'Locomotive soul: the parts of soul in Aristotle's scientific works', *Oxford Studies in Ancient Philosophy* 22: 141–200.

1990. 'Aristotle on form and generation', in J. Cleary and D. Shartin (eds.), *Proceedings of the Boston Area Colloquium in Ancient Philosophy*, pp. 35–63. Lanham: University Press of America.

Williams, C. J. F. 1999. *Philoponus: Aristotle on Coming-to-Be and Perishing.* Translated with an introduction. London: Bloomsbury.

1982. *Aristotle's De Generatione et Corruptione.* Translated with notes. Oxford: Clarendon Press.

Witt, C. 2015. 'In defense of the craft analogy: artifacts and natural teleology', in M. Leunissen (ed.), *Aristotle's Physics: A Critical Guide*, pp. 107–20. Cambridge: Cambridge University Press.

2003. *Ways of Being: Potentiality and Actuality in Aristotle's Metaphysics.* Ithaca: Cornell University Press.

1998. 'Form, normativity, and gender in Aristotle: a feminist perspective', in C. Freeland (ed.), *Feminist Interpretations of Aristotle*, pp. 118–37. University Park: Penn State University Press.

1994. 'The priority of actuality', in T. Scaltas, D. Charles, and M-L. Gill (eds.), *Unity, Identity, and Explanation in Aristotle's Metaphysics*, pp. 215–28. Oxford: Clarendon Press.

1989. *Substance and Essence in Aristotle*. Ithaca: Cornell University Press.

1985. 'Form, reproduction, and inherited characteristics in Aristotle's *Generation of Animals*', *Phronesis* 30(1): 46–57.

Yu, J. 2003. *The Structure of Being in Aristotle's Metaphysics*. Vol. 52, The New Synthesis Historical Library. Dordrecht: Springer.

Index Locorum

Aristotle

Subject Index

Printed in the United Kingdom by TJ International Ltd

Printed in the United Kingdom by TJ Clays Ltd.